AF601530

Reproductive Diversity of Plants

Marines Marli Gniech Karasawa
Editor

Reproductive Diversity of Plants

An Evolutionary Perspective and Genetic Basis

Editor
Marines Marli Gniech Karasawa
Department of Genetics
Luiz de Queiroz College of Agriculture
University of Sao Paulo ESALQ/USP
Piracicaba, SP, Brazil

ISBN 978-3-319-21253-1 ISBN 978-3-319-21254-8 (eBook)
DOI 10.1007/978-3-319-21254-8

Library of Congress Control Number: 2015947777

Springer Cham Heidelberg New York Dordrecht London

Printed on acid-free paper

Springer International Publishing AG Switzerland is part of Springer Science+Business Media (www.springer.com)

I DEDICATE

To God for all the inspiration and strength granted.

To my parents Albano and Teresinha for the life lessons.

To my teachers for the valuable lessons.

To my beloved daughter Eliane for all the encouragement and dedication.

Marines Marli Gniech Karasawa

Preface

The evolution of biology, and the genetics of the diversity of reproductive system in plants is important to understand, conserve, and manage plant genetic resources where the reproductive system plays the major tool in the maintenance of the structure and genetic diversity in natural plant populations.

The *Reproductive Diversity of Plants: An Evolutionary Perspective and Genetic Basis* is divided into two topics. In the first chapter, we present a comprehensive approach of the most important steps on the evolution of plant reproductive systems considering the earth condition at that time. In the second chapter, we discussed about the biology and genetics of the asexual and sexual plant reproductive system.

The book has several unique components, including the treatment of the succession of earth's environmental conditions over the Phanerogamic as well as the biology and the genetics mechanisms underlying plant reproductive systems.

Piracicaba, SP, Brazil — Marines Marli Gniech Karasawa

Acknowledgments

The organizer expresses her gratitude to the professionals who collaborated in this project, reviewing the chapters. Gratitude and appreciation are extended also to the co-authors, to the Department of Genetics of "Luiz de Queiroz" College of Agriculture, University of São Paulo, Brazil, and to Prof. Roland Vencovsky for the ideas and encouragement given to construct this book.

Contents

Contributors

A.C.G. de Araújo EMBRAPA—Recursos Genéticos e Biotecnologia, Prédio de Biotecnologia, Parque Estação Biológica—PqEB, Brasília, DF, Brazil

M.C. Dornelas Department of Plant Biology, Biology Institute, State University of Campinas, Campinas, SP, Brazil

M.M. Gniech Karasawa Department of Genetics, Luiz de Queiroz College of Agriculture, University of Sao Paulo ESALQ/USP, Piracicaba, SP, Brazil

G.C.X. Oliveira Department of Genetics, Luiz de Queiroz College of Agriculture, University of Sao Paulo ESALQ/USP, Piracicaba, SP, Brazil

E.A. Veasey Department of Genetics, Luiz de Queiroz College of Agriculture, University of Sao Paulo ESALQ/USP, Piracicaba, SP, Brazil

Chapter 1
Evolution of Plants with Emphasis on Its Reproduction Form

Marines Marli Gniech Karasawa, Giancarlo Conde Xavier Oliveira, and Elizabeth Ann Veasey

Abstract Chapter 1 of this book discusses the evolution of plants from the Pre-Cambrian era considering the starting time of alternation of generations. In the Paleozoic and Mesozoic eras, we will discuss the transitions from homospory to heterospory and the formation of the egg cell and the pollen grain as well as the evolution of seed plants. At the end of this chapter we will discuss the evolution of the angiosperms (flowering plants) and the sexuality transitions and the reproductive strategies. Also, the evolution of the unisexuality, self-incompatibility systems, selfing, and mixed mating systems will be discussed considering their evolutionary implications.

1.1 First Plants and the Alternation of Generations

The fossils of the first organisms that performed photosynthesis and produced oxygen were found in Warrawoona, Western Australia. These spherical and filamentous microorganisms with 3300 and 3500 million years (myr) were classified as photoautotrophic cyanobacteria (blue-green algae) (Schopf and Packer 1987). Further evidence of the evolution of plants from photosynthetic unicellular organisms that divide by mitosis was found in deposits in South Africa. These organisms, belonging to the Pre-Cambrian period (Table 1.1), were dated between 3200 and 3100 myr (Freeman and Herron 1998; Brown and Lomolino 2005). In more recent sediments of rocks of Southern Ontario, Canada, with approximately 2000 myr, specimens of blue-green algae were also found. These early ancestors of plants showed prokaryotic life because they did not have organized nuclei as in higher organisms. Thus, the oldest and the most reliable trace available from the emergence of eukaryotes has an age between 2000 myr (Mussa 2004) and 1400 myr (Zunino and Zullini

M.M. Gniech Karasawa (✉) • G.C.X. Oliveira • E.A. Veasey
Department of Genetics, "Luiz de Queiroz" College of Agriculture, University of Sao Paulo, Avenida Pádua dias, 11, Piracicaba, SP, 13418-900, Brazil
e-mail: mgniechk@gmail.com

M.M. Gniech Karasawa (ed.), *Reproductive Diversity of Plants*,
DOI 10.1007/978-3-319-21254-8_1

2003), when the first members of unicellular organisms called acritarchs, belonging to the Protist kingdom, appeared in the fossils record in China. It is widely accepted that the ancestors of land plants are in the *Charales* order of the Coleochaete genus (Graham et al. 2000). Protists were commonly found in aquatic environments, but some were able to live in terrestrial environments. Generally they had sexual reproduction, but some could reproduce asexually (Raven et al. 2007). Analyses of rock sediments with about 1000 myr revealed a wide variety of types of algae, the vast majority bluish-green. Apparently they had a true nucleus and mitotic divisions (Fig. 1.1), being among the first eucaryotic organisms that we know of so far (Schopf 1968; Banks 1970; Knoll 1992).

In response to evolutionary changes, living organisms became increasingly diverse and complex in structure (Raven et al. 2007). These organisms have evolved for about another 500 myr until the first plants appeared with multicellular and erect growth. Geological evidence suggests the green algae *Fritschiella* as a probable ancestor of land plants, because metabolism features are similar to that of existing plants, which was not found in any other algae (Fig. 1.2). *Fritschiella* lived in fresh water, but could be found in any other terrestrial environment (Willis and McElwain 2002).

In the Cambrian and Ordovician periods (Table 1.1) tectonic activities were relatively intense promoting the reorganization of the continental plates deeply interfering in the sea levels, ocean currents, climate and geographical distribution of

Table 1.1 Geological time scale with major evolutionary changes

Era	Period	Major events in the evolution on plant kingdom	Years (myr)
Cenozoic	Neogene		
	Paleogene		23.8
Mesozoic	Cretaceous	Evolution of angiosperms	65
	Jurassic		144
	Triassic		206
Paleozoic	Permian	Major expansion of seed plants	248
	Carboniferous		290
	Devonian		354
	Silurian	Evolution of vascular land plants	417
	Ordovician		443
	Cambrian		490
			543
Pre-Cambrian		The first algae	4600

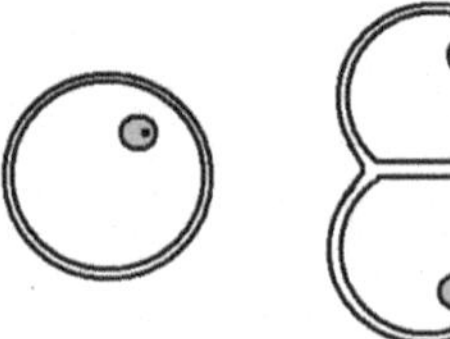
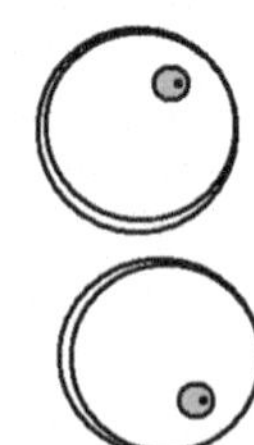

Fig. 1.1 Schematic representation of the division by mitosis in unicellular algae (Karasawa et al. 2006)

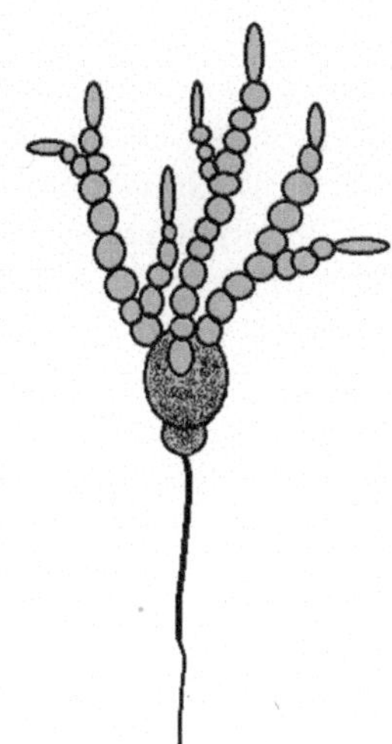

Fig. 1.2 Diagram of the probable ancestor of land plants (Drawn by MMG Karasawa based on Willis and McElwain 2002)

organisms affecting the life of all organisms of that time. Evidence suggests that in the high Cambrian all continents were distributed around the equator. The largest Gondwana moved to the South Pole while the two smaller coalesced into a continental landmass called Laurasia (Zunino and Zullini 2003). The reorganization of continental plates promoted the occurrence of glaciation (440 myr) that led to a dramatic reduction of sea level of approximately 70 m and around this period the first evidences of land colonization were found (Brown and Lomolino 2005; Lomolino et al. 2006). In parallel to changes in environmental conditions, changes were necessary in the structure, form and reproduction of plants to ensure their existence in the terrestrial environment (Willis and McElwain 2002). At the end of the Cambrian period green algae had evolved into highly complex shapes. But, the biochemical pathways, such as those that facilitated respiration and photosynthesis in the cyanobacteria, fundamental to the life of plants and algae who occupied aquatic environments, and the advent of meiosis, that promoted the emergence of more sophisticated forms of life, were established only in the Ordovician period (510~438 myr) (Bateman et al. 1998). Among the fossil specimens recorded, the best preserved is the *Isochadites* of Codiaceae family. This fossil shows gametocysts with reproductive structures, probably one of the earliest evidences of sexual reproduction (Banks 1970). The main algal groups from the Cambrian period were *Dasycladaceae* and *Codiaceae*, among the green algae, and *Solenoporaceae*, among the red algae. In the intermediate stage of the Ordovician period the *Codiaceae* presented segmented structures and an internal tubular structure found until today.

From the middle of the Ordovician period to the beginning of the Silurian (470~430 myr) (Table 1.1), fossils evidence were found from the development of specialized cells for water and nutrients transport, such as some other precautionary measures against desiccation, mechanical support and reproduction mode that decreased environmental water dependence (Willis and McElwain 2002). The aerial parts and the underground sporophytes of the first vascular plants differed little structurally one from another, but undoubtedly the primitive plants resulted in more specialized plants with differentiated body. These plants consisted of roots, that functioned in the fixation and absorption of water and minerals, and stems and leaves, which provided a well-adapted system to the necessities of life on earth (i.e., absorption of sun light, carbon dioxide from the atmosphere and water from the soil) (Raven et al. 1995, 2007). The Cooksonia *Aglaophyton major*, also known as *Rhynia major* (Edwards 1986), can be considered an intermediate stage in the evolution between primitive and vascular plants (Prototracheophyte) because they do not have tracheids. However, the presence of cells similar to moss hydroids was detected. During the transition, plants also underwent other changes that made possible their reproduction in terrestrial environments, with the production of resistant spores being one of the earliest stages for enduring drought (Raven et al. 2007). As an example, one can cite Cooksonias (Fig. 1.3) which were formed by sporangia containing spores inside (Mussa 2004). These sporangia could reach a maximum of 1.5 mm in diameter and 2.0 mm in length. The height of these plants ranged from 2.2 cm to 11 cm (Edwards et al. 2004).

Plant evolution is associated with the occurrence of several changes in the gametophyte and the sporophyte (Fig. 1.4a). Currently, the most accepted hypothesis is that a spore producing ancestor, which lived in aquatic environments, has given rise to the first plants (Fig. 1.4a). This would have changed (by mutations) in gametophytic and sporophytic phases, some of which resulting in plants with an amplified gametophytic generation, which was nutritionally dependent (Fig. 1.4b) and presented the reproduction mode of most non-vascular plants (bryophytes). Some other changes resulted in plants with an amplified sporophytic generation, who produced nutritionally independent sporophytes (Fig. 1.4c), with a reproduction mode similar to that of vascular plants (tracheophytes) (Willis and McElwain 2002). The emergence of the gametophytic stage must have kept the dependence on water, which has become essential to transfer male and female gametes, as well as for the initial growth of the sporophytic embryo. Furthermore, plants that have developed an amplified sporophytic generation showed a decrease in water requirement as a consequence of a continuous drying of the environment, and were selected for a nutritionally independent sporophytic stage. Thereby, neither the spore production nor

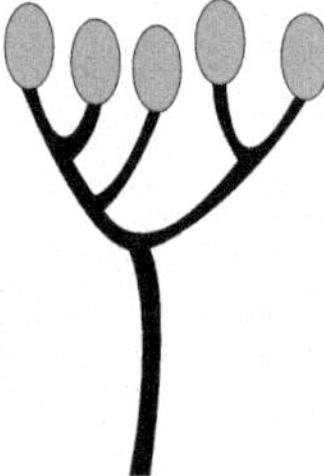

Fig. 1.3 Diagram showing morphology of Cooksonia (Drawn by MMG Karasawa)

a) Hypothetical algal precursor of terrestrial plants

Diploid sporophyte (2n)

Zygotes germinate to form diploid plants called sporophytes

Meiosis

Free-swimming haploid spores (n)

Syngamy (gamete fusion) in water

Male and female gametes fuse to form zygotes

Haploid gametophyte (2n)

b) Bryophyte

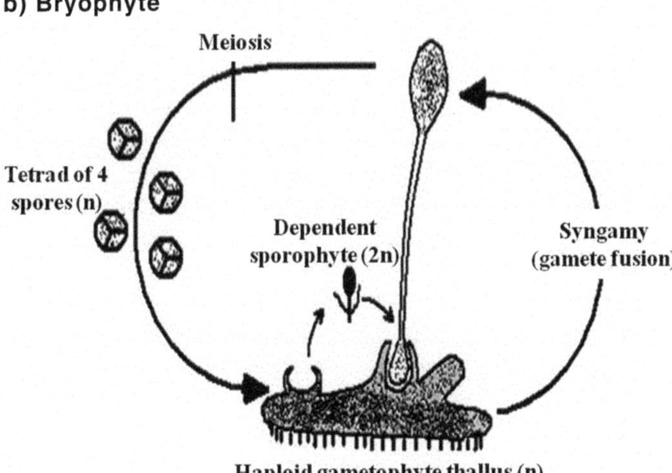

Fig. 1.4 Diagram showing the alternation of generations with (**a**) gametophytic and sporophytic phases in algae and (**b**) amplified gametophyte generation as in most bryophytes. Plants of simplified life cycle with (**c**) amplified sporophytic generation that occurs in all vascular plants (**a**) hypothetical algal precursor to terrestrial plants (**b**) bryophyte (**c**) tracheophyte. Drawn by Karasawa et al. 2006

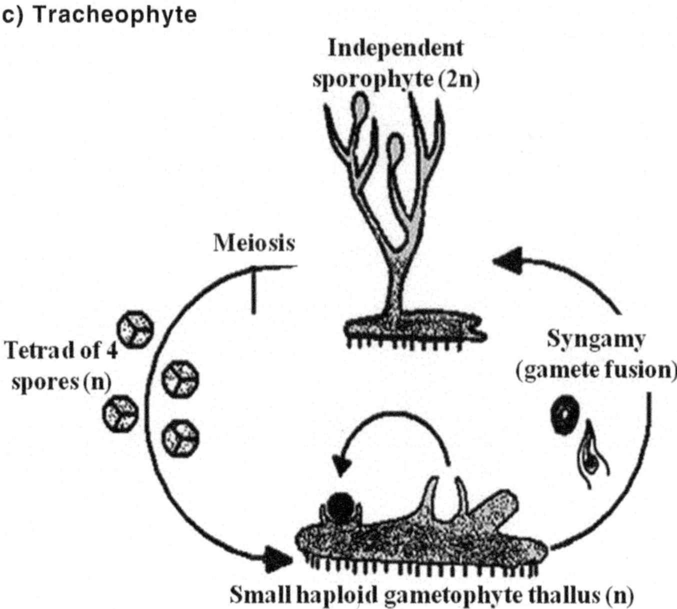

Fig. 1.4 (continued)

its spread needed water. Mutant individuals that arose in this period showed reduction in size and in gametophytes complexity compared to sporophytes (Fig. 1.4c) and gave rise to an independent sporophytic phase. This strategy increased the resistance to drought and dissection. Other mutants presented a larger gametophyte size and became physiologically dependent on moist environments to transfer their gametes (Fig. 1.4b) due to the amplified gametophytic generation (Drews and Ydegari 2002; Graham et al. 2000).

Many algae groups are known to reproduce sexually and asexually, and sexual reproduction surrounds the alternation of two kinds of generations, viz., the diploid sporophytic and the haploid gametophytic phases. In the gametophytic phase, male and female gametes are released from the gametophyte and the male gamete swim to the female gamete. They then merge to produce a diploid zygote. The germination of the zygote form plants called sporophytes that, when mature, undergo meiotic division to form haploid spores that are released to form a new gametophyte.

The differentiation between the amplification of either the gametophytic or the sporophytic generation has persisted to the present. In the current vascular plants the vegetative sporophyte is the visible part of the plant (Figs. 1.4c and 1.5), whereas in the non-vascular group the vegetative gametophyte is the visible part of the plant (Figs. 1.4b and 1.5). This differentiation may be the explanation why, throughout the geological record, bryophytes have remained small and restricted to humid environments and mixed areas, while the evolution of tracheophytes not only turned them into the largest group of plants in the planet, but also allowed them to occupy the most diverse ecological niches.

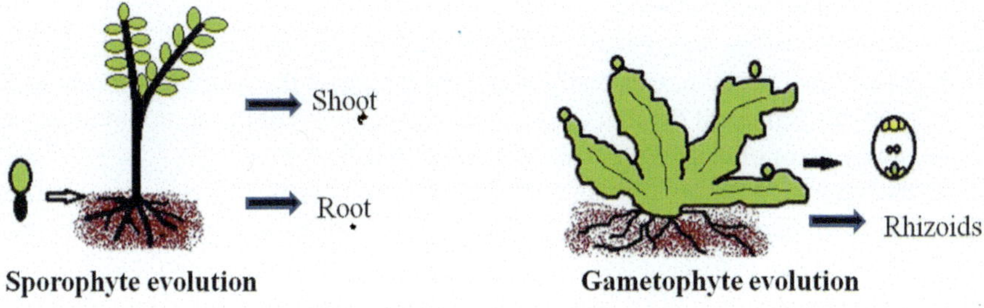

Fig. 1.5 Sporophyte and gametophyte evolution (Drawn by MMG Karasawa)

Fig. 1.6 Spores arranged in tetrahedron form (**a**); composition of tetrahedron of spores (**b**, **c**) and isolated spores with distinct trilete form (Drawn by MMG Karasawa based on Willis and McElwain 2002)

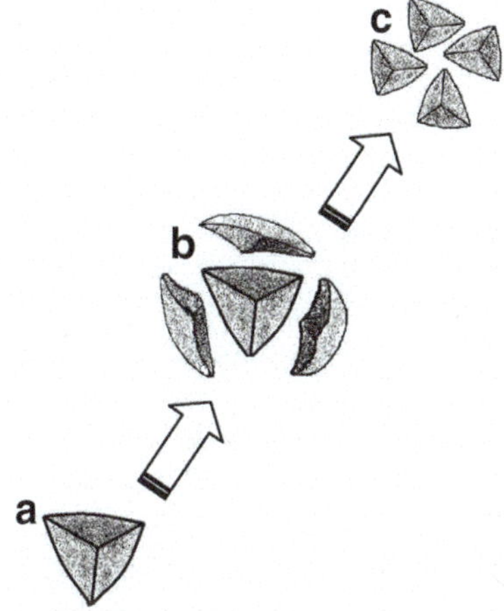

It has been suggested that the appearance and gradual increase in the number of spores in the fossil record has not only been an indication of the sporophytic phase development, but also of the fact that they have become resistant to deterioration, because of the presence of sporopollenin in the wall. This chemical substance can be found in the wall of the pollen grain of non-vascular plants, vascular plants and also in some algae (Kenrick and Crane 1997). This complex polymer has the function of providing drought resistance, strength and protection against ultraviolet radiation. Evidence from fossil spores indicate that land plants were originated in the intermediate stage of the Ordovician period, while the greatest divergence between hepatics, ceratophyllum, mosses and vascular plants groups must have occurred at the end of the Ordovician and in the Silurian (Bateman et al. 1998). Some of the early visible spores found in the geological records belonged to the end of the Ordovician period (~450 myr). They had a tetrahedron arrangement while others from younger sediments (~430 myr) are found as isolated spores with a distinct trilete form (Gray and Shear 1992). This trilete morphology (Fig. 1.6) as well as the tetrahedron arrangement provides strong evidences for the meiotic division because a diploid cell, when dividing by meiosis, produces four haploid cells (in this case, spores). The significance of the fossil spores arranged with a tetrahedron form is that most of them, especially in the beginning of the Silurian period (~430 myr), probably represent the first evidence of development of the sporophytic phase in the life cycle of plants (Willis and McElwain 2002).

At the beginning of the Silurian period (~417 myr) fossil evidence was found of the earliest vascular plants. They presented globose sporangia with cutinized spores and isolated spores in trilete form. Multiple reproductive structures were also found. The growth habit of these plants was the determinate type (i.e., terminal reproductive structures), and they had asexual spreading rhizomes and/or sexual reproduction with spores (Banks 1970).

1.2 Homospory, Heterospory and the Evolution of the Egg and the Pollen Grain

Until the beginning of the Devonian period, the fossil records show only the presence of determinate growth habit. The first evidence of plant growth habit of indeterminate type comes afterwards. In the beginning of the Devonian, the *Rhynie* type is one of the most important evidences of growth habit and reproductive structures. Their fossil records show various forms of reproductive structures found in plants that may be isolated, multiple and even fused (Willis and McElwain 2002; Mussa 2004).

Evolution promoted the emergence of new types of plants and the extinction of old types in the Devonian period. The blue-green algae reached their peak in the beginning of this period, while the Characeae were found in more advanced stages of evolution. These algae inhabited fresh water, but occasionally were found in seawater, and can be recognized by the arrangement of its branches with a structure characterized by nodes and internodes, and their sexual organs, the oogonia, which was attached to the nodes. The oogonia was a single egg cell surrounded by coiled tubules. The most ancestral fossil of this group is represented only by its zygote. These individuals became highly specialized in the early Devonian period, and these structures have been conserved until today, differing only in some minor details (Banks 1970). Another marine algae group, also found in this period, was the Dasycladaceae and the Codiaceae, inhabiting environments that contained fresh water and also brackish water. It is believed that this group would have migrated from marine environments to fresh water, at the end of the Silurian period, because its zygote (oospore) was able to resist to desiccation, where the water supply could dry occasionally, that was an obvious adaptation to the new habitat. This type of spore resistance was not a common characteristic between seawater algae.

Land plants and Charophyceae algae may have moved to new environments at the same time in the past. According to the existing hypothesis, all plants have originated in marine environments wherein the migration to terrestrial environment would have led to new ecological niches. It has been reported, between the fossil samples of the Rhynie type, as found in the Devonian period, *Zosterophyllum divaricatum* with 400 myr, with sporangia attached laterally and at the apex of the stem (Gensel and Andrews 1987). Their spores were small, with a diameter of approximately 55–85 μm, ranging from circular to subtriangular, smooth and with a distinct trilete mark. This plant was approximately 30 cm in height and was composed of branches that grow from rhizomes. However, *Psilophyton dawsonii* (395 myr) and *P. robustius* had a central stem with undeterminate growth which grew to a height of 2–60 cm and had lateral branches with fertile apices, each apex consisting of approximately 32 sporangia (Willis and McElwain 2002; Mussa 2004).

The first representatives of vascular plants were the moss group, and among them three genera, *Asteroxylon*, *Calpodexylon* and *Protolepidodendron*, illustrate similar characteristics between the modern and the fossil type. In this phase (~400 myr) the plants produced globose and reniform sporangia with spores in the trilete form that, during this period, evolved from homospory (spores with the same size) (Fig. 1.7a) to heterospory (spores with different sizes) (Fig. 1.7b, c), in which the small spores are called microspores (3~5 μm) and the large spores are called megaspores (150~200 μm) (Mussa 2004). This is considered one of the most important courses of evolution for the emergence of seed plants (the gymnosperms) (Willis and McElwain 2002).

It is postulated that the largest spores were the precursors of today's megaspores, and the smallest were the precursors of pollen grains. The most accepted theory is that a mutation would have given rise to two sizes of spores (Fig. 1.8) (Thomas and Spiecer 1987).

These spores of different sizes would have been initially developed together in the same sporangia (Fig. 1.7b) and, subsequently, over the course of evolution, would have arisen in separated sporangia (Fig. 1.7c), the megaspore in the megasporangia and the microspores in the microsporangium (Andrews 1963; Banks 1970).

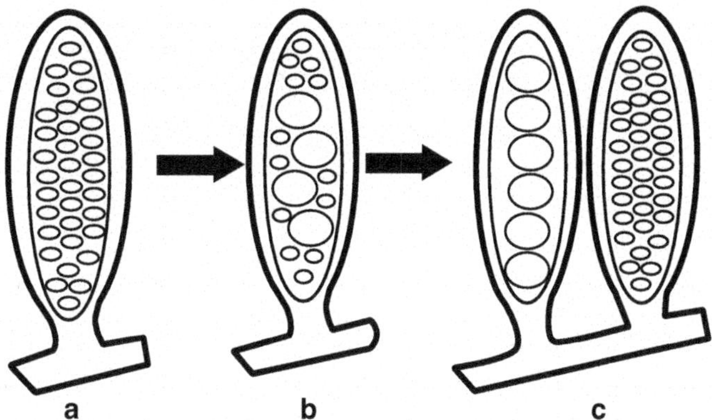

Fig. 1.7 Sporangium with homospores (**a**), sporangium with heterospores (**b**) and two sporangia on the same individual, one bearing only megaspores and the other bearing only microspores (**c**) (Karasawa et al. 2006)

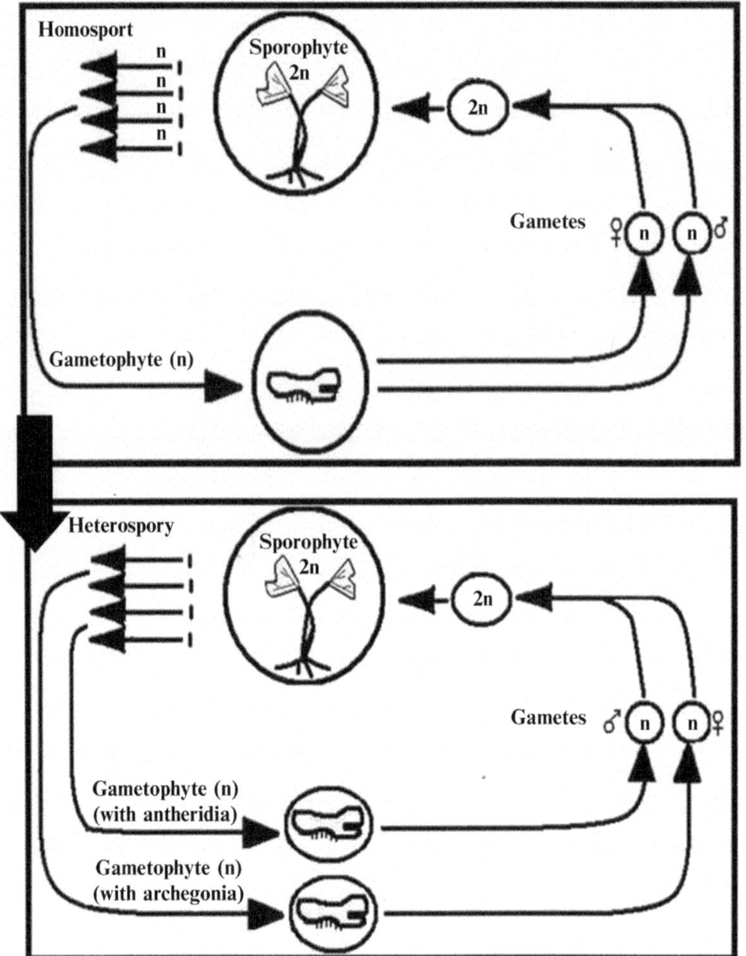

Fig. 1.8 Diagram illustrating the transition from homospory to heterospory in vascular plants (Willis and McElwain 2002, modified by Karasawa et al. 2006)

According to the fossil records, heterospory gradually evolved in the Devonian period. The evolution of megasporogenesis led eventually to the degeneration of three out of four products of female meiosis, which produces now only one viable megaspore per megasporocyte (Fig. 1.9a–c). The surviving megaspore can thus receive more nutrients from the mother plant (Fig. 1.9d) (Willis and McElwain 2002).

But this megaspore was still subject to attack and desiccation, so that natural selection soon led to the evolution of fused sterile leaves located nearby for its protection, as shown in Fig. 1.10 (Thomas and Spiecer 1987).

It is believed that this development would have occurred around 370~354 myr with the evolution of the pro-gymnosperms (Willis and McElwain 2002). Support

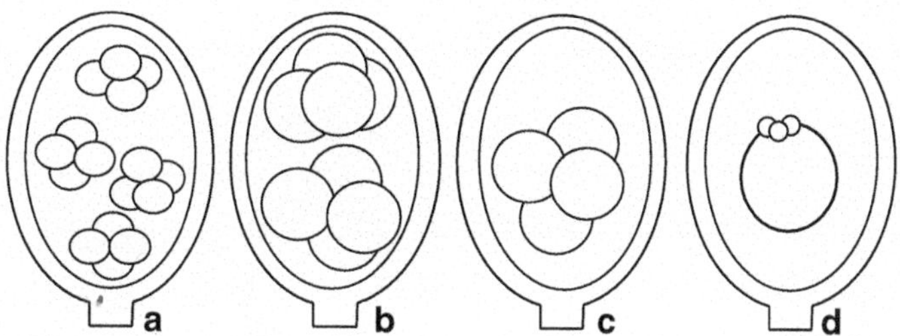

Fig. 1.9 Scheme illustrating the degeneration of megaspore indicating the probable evolution of the megasporogenesis in the megasporangium (Willis and McElwain 2002, modified by Karasawa et al. 2006)

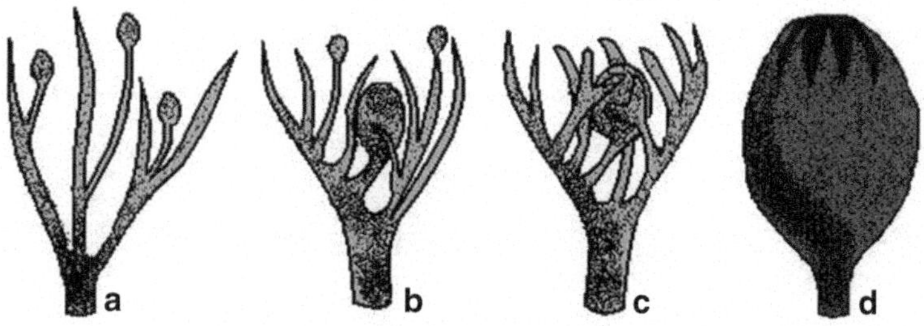

Fig. 1.10 Stages in the evolution of megaspore protecting structures from sterile leaves existing nearby (Karasawa et al. 2006 based on Andrews 1963 and Willis and McElwain 2002)

for this theory is provided by *Genomosperma kindstoni*, *Genomosperma latens* and *Salpingostoma dasu* (early Carboniferous), and *Physostoma elegans*, *Eurystoma angulare* and *Stamnostoma huttense* (late Carboniferous) (Andrews 1963).

The megaspore and the microspore evolved at the same time. The microspore has given rise to the pollen grain that could travel a long distance and was able to develop a pollen tube to reach the female gamete into the embryo sac, where the zygote is formed. Paleobotanical evidence shows that the pollen grain began its evolution around 364 myr, and only fossil spores are found before that. The first evidence of the pollen grain shown in the fossil record was termed pre-pollen and corresponds to an intermediate stage between the spores and the pollen grain, while it contained characteristics of spore (trilete form), but the evidence suggested that the germination took place on or near the opening of the megasporangium (Fig. 1.11) (Willis and McElwain 2002).

The pollen grain is distinguished from the spore in the structure and in the mode of gamete release. In heterosporic plants, the microspores release flagellated gametes at the spore distal end (i.e., the opening of the trilete), which swim to the archegonium for fertilization. The pollen grain, by comparison, produces a pollen tube at the distal end through which the gametes are transferred directly to the egg (Fig. 1.12).

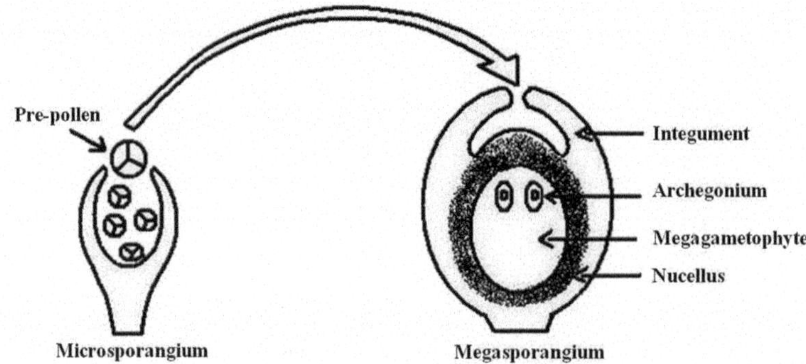

Fig. 1.11 Scheme illustrating the likely mode of pollination used by the pre-pollen to reach the pre-egg (Karasawa et al. 2006)

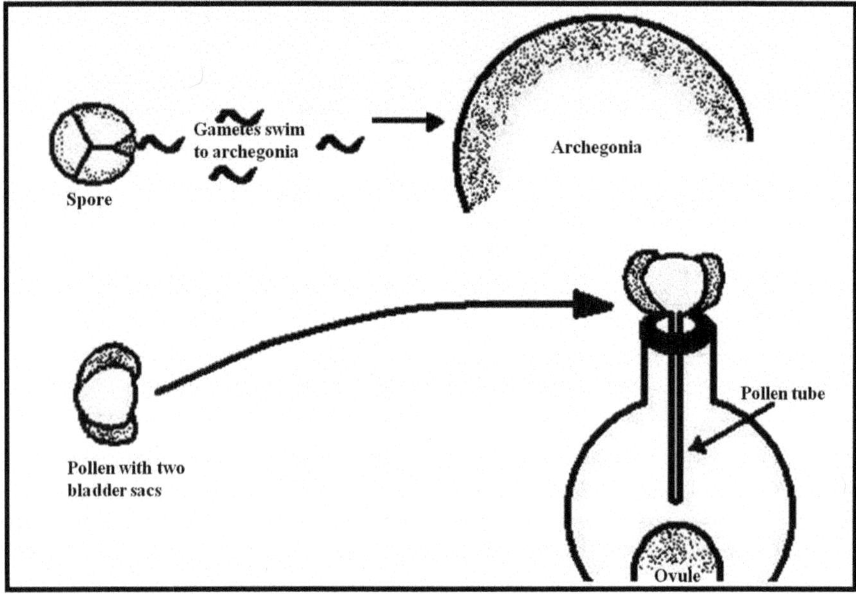

Fig. 1.12 Diagram showing the difference in the release of the gametes from spores and from pollen grains (Willis and McElwain 2002, modified by Karasawa et al. 2006)

With the development of seed protection, it has become necessary to improve the mechanism of pollen reception by the gynoecium so that it could reach the ovule in the ovary to form the zygote by fertilization. In the first seed plants the egg protection was partial, and the sterile leaves were fused only at the megasporangium basis (complete fusion later would evolve producing the carpels), leaving the structure that received the pollen free (Fig. 1.13a). These structures, combined with the protection lobes, were highly effective in capturing the pollen grain carried by the wind. However, some of the first eggs had other mechanisms to receive pollen, such

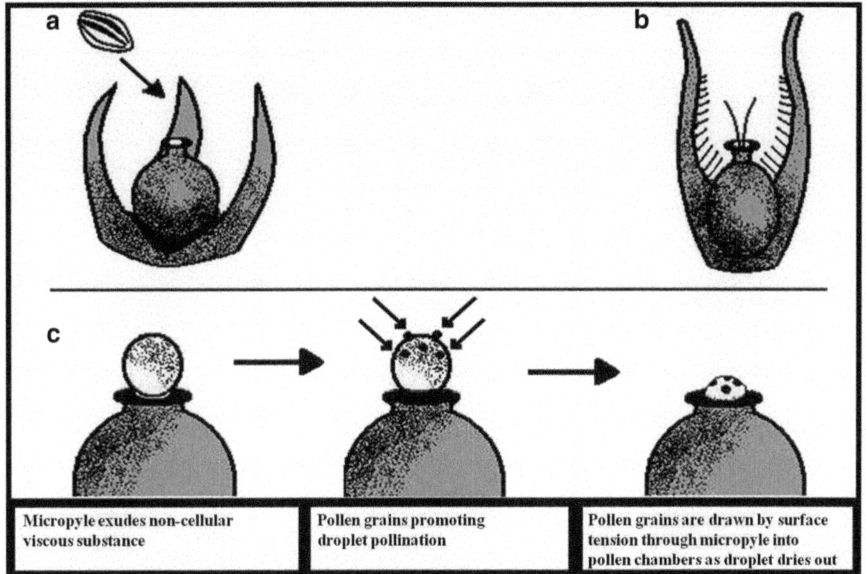

Fig. 1.13 Mechanisms of pollen reception present in the fossils. (**a**) Presence of lobes, (**b**) inner and outer fringes, and (**c**) droplet pollination (Willis and McElwain 2002, modified by Karasawa et al. 2006)

as the presence of inner and outer fringes in the integumentary lobes (Fig. 1.13b) and the droplet pollination (Fig. 1.13c). The droplet pollination mechanism used a substance consisting of an exudate, which adhered to the pollen after their deposition on the surface, until the dehydration and consequent volume reduction brought the pollen grain inside the micropyle allowing fertilization (Thomas and Spiecer 1987; Willis and McElwain 2002).

1.3 The Evolution of the Terrestrial Flora

From the end of the Devonian to the beginning of the Permian period (395~286 myr), the terrestrial flora evolved from small non-vascular and vascular plants to vegetation that included trees with 35 m or more. Due to adaptations to live in terrestrial environment, vascular plants have been ecologically successful, becoming numerous and diversified in the Devonian period (Willis and McElwain 2002; Raven et al. 2007). During this phase there was a significant change in the global environment, as the movement of continental plates promoted a dramatic change in the climate. Under these conditions major innovations occurred in the morphology of terrestrial vegetation, with the emergence of new plants (390~365 myr) and a relatively rapid increase in the number of species (Gensel and Andrews 1987; Lomolino et al. 2006). Everything indicates that the environmental selection pressure was probably the main factor in these events. The fossil records of that time suggest a steady increase in ecological complexity at all spatial scales. Ecosystems, from the beginning of the Devonian, where composed of plants with simple and dynamic interactions. Differences in local dynamic in landscape scales were small and difficult to differentiate. The communities consisted of groups of plants with clonal and opportunistic reproduction (Willis and McElwain 2002). Two are the groups of plants known to have evolved at this early period: *Sciadophyton* and *Protobarinophyton* (Banks 1970).

Typical plants with rhizoids and rudimentary roots co-evolved, supporting the turgor pressure and showing a history of evolution of homospory. The best known history is that of *Rhynie*, which provides crucial informations about the ecosystem. Yet sporophytic structure was simple; many sporophyte ecological strategies clearly co-existed, as the hability of *Rhynie gwynne-vaughanii* to disperse rapidly on the substrate by lateral and decidual branches. The diversification of the sporophyte was exchanged by a wide array in the morphology of the gametophyte which registers the many variations helping the syngamy. In the mid-Devonian period precursors of modern horsetail (Sphenopsida) were found. Two other extinct types, *Cladoxylon* and *Aneurophyton*, were also found. An empirical study of megafloras and deposition environments has shown the occurrence of early stages of landscape partition by a group of higher plants. Typical floras of wetlands were dominated by a genus with fern features (*Rhacophyton*), and the adjacent areas included the lycopods, which were distinguished from the interfluve floras, while the dry parts of the plains were dominated by the pro-gymnosperm *Archaeopteris* (Willis and McElwain 2002).

Also, at the end of this period (370~354 myr) the evolution of the ovule occurred giving rise to the seed, which was one of the most spectacular innovations that emerged during the evolution of vascular plants. The emergence of the seeds was one of the major factors responsible for the dominance of the current seed plants, which evolved over a period of several hundred million years. The main factor of this success was the fact that the seed gives the embryo stored food that becomes available in critical stages of germination and establishment, thus promoting a selective advantage in relation to related groups endowed with free spores and ancestral groups that liberated spores (Raven et al. 2007).

Apparently, the Pteridosperms, first seed plants, originated in humid landscape and then settled as opportunistic in disturbed landscapes with physical stress, including the relatively arid habitats. Fossil evidence showed that at the same time the main stem structure, that characterizes the most modern groups of gymnosperms, ferns, Sphenophyta and a series of lycopod groups were developed (Willis and McElwain 2002). When forming the structure of the main stem, the development of the conducting system of the central cylinder (the eustele) was a major novelty which allowed a greater growth in height and the effective transport of water and nutrients throughout the plant up to the canopy (Raven et al. 2007). Geological evidence suggests that the conductive system became progressively more complex over the evolution of vascular plants, apparently presenting, at the end of the Devonian period (~374 myr), three different types namely: protostele, siphonostele and eustele (Fig. 1.14a–c).

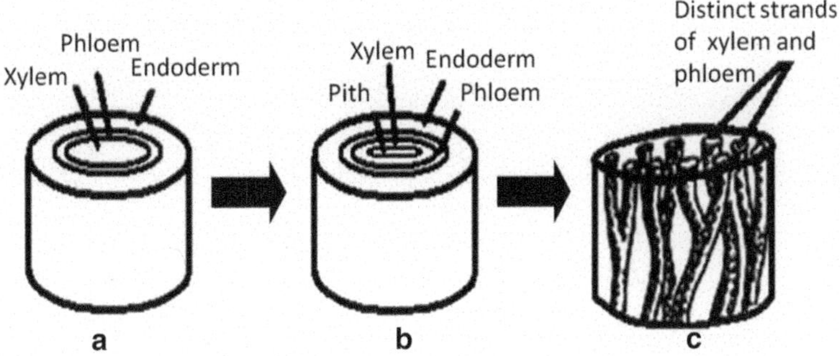

Fig. 1.14 Morphological differences between the fossils of the earliest eustele types. (**a**) protostele: vascular plants ~420 myr; (**b**) siphonostele: fossils with about 395 myr; (**c**) eustele: fossils with about 380 myr (Karasawa et al. 2006)

According to the fossil record, lycopods, sphenopsida, filicinaceae and pro-gymnosperms were the first trees producing spores. The ancestral lycopods were small plants, herbaceous and homosporous, as for example the *Baragwanathia longifolia*, with approximately 410 myr, while the first trees were found in fossils dating back to 370 myr, with the giant species *Lepidodendron* being one of the most commonest among the lycopods. It reached a height of 10~35 m and a diameter of 1 m. This species was heterosporous and the sporophylls were encountered in strobili containing microspores at the top and megaspores at the bottom (Willis and McElwain 2002; Mussa 2004).

The Sphenophyta currently comprises a group of 20 species, all having herbaceous growth habit and belonging to a single genus. Many fossils of this group have been assigned to the Carboniferous and Permian period (354~248 myr), including a series of arborescent forms, most of them belonging to *Calamites*, which grew to 18 ft of tall height or more. This plant had siphonostele stem with primary and secondary xylem and primary phloem. Another striking feature of the stem was the presence of nodes. Regarding the spore type they were homosporous. However, strong evidence point to the presence of heterospory in this group at the end of the Carboniferous (Willis and McElwain 2002).

The Filicinaceae can be found in the fossil record of about 360 myr, and many fossils in this group are quite similar to the present remaining forms. The *Psaronius* is an example of fossil of this group. It grew to a height of about 10 m, had long leaves and a protostele type of stem (Thomas and Spiecer 1987). Paleobotanical evidence suggests that in some species the root reached 1 m of diameter at the base of the stem. Most plants of this group were homosporous. In *Psaronius* the sporangium was big and had fused locus pairs (synangium). The bottom branches of the cones were fertile, suggesting that the arrangement evolved with the connecting to the megaphylls incorporating the sporangia at the bottom (Willis and McElwain 2002; Mussa 2004).

1.4 Evolution of Pro-gymnosperms and Gymnosperms

The second mass extinction occurred in the period between 395 and 290 myr. This event was marked by significant changes in the global environment accompanying the collision of the blocks that formed Gondwana and Laurasia during the Silurian, which formed the supercontinent Pangaea. The active movement of tectonic plates promoted a dramatic change in the climate, which changed from hot (24–32 °C—near the equator) to cold, with very low temperatures and dry climate (in the inner of the continent), and there were extensive glaciations in high regions in the southern hemisphere, with more intense effects in the inner part of the continent (Freeman and Herron 1998). During this period the sea level dropped 100–200 m. Moreover, land colonization and the consequent atmosphere CO_2 reduction contributed overwhelmingly to global cooling (Zunino and Zullini 2003; Lomolino et al. 2006).

The evolution of pro-gymnosperms occurred between the end of the Devonian period and the early Carboniferous period (~354 myr). The pro-gymnosperms comprise a group of plants representing the transition between ferns and gymnosperms. This group had some type of secondary xylem and phloem, the presence of bifacial vascular cambium and determined growth habit in some plants and indeterminate in others. In these, the most important advance, in relation to their ancestors, consisted in the fact that they presented a bifacial vascular cambium (i.e., a vascular cambium that produced secondary xylem and phloem). This type of vascular cambium is characteristic of seed plants and, apparently, first evolved in pro-gymnosperms. Among the pro-gymnosperms from the Devonian period, *Aneurophyton* (380~360 myr), a plant characterized by having tridimensional complex branching and protostele (i.e., a closed cylinder of vascular tissue), may be mentioned. Another important pro-gymnosperm was the *Archaeopteris* (370~340 myr) (Banks 1970). This plant, which also lived in the Devonian, had lateral branches with flat laminar structures considered leaves and presented stems of the eustele type (with single strands of vascular tissue arranged around the pith).

Regarding the reproductive system, most of the pro-gymnosperms were homosporous, but some species of *Archaeopteris* were heterosporous. Several groups of vascular plants without seeds thrived during the Devonian period, in which three of the most important were: Rhyniophyta, Zosterophyllophyta and Trimerophytophyta. These three phyla consisted of plants without seeds that presented a relatively simple structure, and all came to extinction at the end of the Devonian, approximately 360 myr. Only a fourth phylum of vascular plants without seeds, Progymnospermophyta, with intermediate characteristics between vascular trimerophyte without seeds and seed plants, did not become extinct. It is speculated that this phylum was the ancestor of the seed plants, the gymnosperms and angiosperms. Although these plants reproduced freely by the dispersal of spores, they produced a secondary xylem (wood) remarkably similar to extant conifers, being the only ones in the Devonian that produced a secondary phloem. The pro-gymnosperms and ferns probably originated from the older trimerophytes (*Rhynia*, *Zosterophyllum* and *Trimerophyta*), from which they differed primarily in having more elaborated and differentiated branch systems and more complex vascular systems than their ancestors (Raven et al. 2007).

With the decline of the seedless plant groups (spore-producing trees), only the Filicinaceae remained as today's remnants of the Paleozoic era. The seed-producing plants appeared on the upper Devonian (~350 myr) and came to dominate the flora of the Mesozoic terrestrial landscape. The seed-producing plants were composed of five classes: pteridosperms (extinct), pteridophytes, cycads, ginkgoales and conifers. The first four had their peak at the end of the Paleozoic and early Mezozoic era. These differ considerably in the structure and form from their ancestors. The seeds of these plants were exposed in the same way as they are in the current pine strobilus (Andrews 1963; McAlester 1978).

The gymnosperms had great advantage over their ancestors because they were able to reproduce without external moisture. The male gametes did not need to swim to fertilize the female gametes, because they were able to form a pollen tube that carried the male gametes to the egg cell allowing the fertilization. The male gametophytic stage no longer required a liquid medium, because the pollen developed within a humid sporophytic plant tissue and the female gametophytic phase, reduced to the embryo sac, was embedded in the sporophyte. The pollen grain surrounded by a double impermeable membrane was highly effective in preventing water loss, and the thinner and more elastic inner membrane gave rise to the pollen tube. The small size and the large number of pollen grains allowed their transportation by wind to the stigma by issuing the pollen tube to reach the egg and promoted fertilization. The wind also made possible for plants that were far apart and from different individuals to become fertilized. After fertilization, the formed seed received all the nutrients (proteins, fats, starch, etc) which helped on the establishment of the embryo in the early stage of its development (Raven et al. 2007).

The pteridosperms was a group of seed plants that showed great development since the lower Carboniferous period. These comprised one of the principal groups of plants of coal-forming plants (McAlester 1978). On the other hand, Cycads and Ginkgoales dominated the landscape of the Triassic and Jurassic period, but declined rapidly during the Cretaceous, as the angiosperms developed, being relatively rare nowadays. It was in response to an increase in temperature and a decrease in moisture in the continent that the evolution of Cycads, Bennettitales and Ginkgoales occurred. These three groups still have living representatives, but in the Mesozoic era its global distribution was much more expressive (Willis and McElwain 2002).

Cycads belong to the group which currently comprises 10 genera and 100 species of plants, all dioecious (i.e., the population consists of plants with male strobili and plants with female strobili) (Mussa 2004), with no fossil record indicating the presence of monoecy (i.e., plants with separate sexes in cones of the same plant). Some of the earliest records are dated from approximately 280 myr (from Permian) and indicate that some species grew up to 15 m in height, although the first cycads were smaller, reaching about 3 m. The apices of these plants have been highly conserved through evolution, being very close to those found in the current cycads. Their reproductive organs were well documented in the fossil record. It is known that in these plants the female reproductive structure had eggs grouped into modified leaves called megasporophylls, and the male structure was located in modified leaves called microsporophylls, wherein each leave had small and compact pollen sacs attached on its surface. In groups of ancestral plants each pollen sac was able to produce a great number of pollen grains in the monolete form (Willis and McElwain 2002).

The Bennettitales, in turn, have fossil records dating from the beginning of the Triassic to the end of the Cretaceous (280~140 myr). This group had many morphological similarities to that of current cycads and also of the extinct ones. One of the genera of Bennettitales most commonly cited is *Williamsonia* (Banks 1970). This plant had reproductive structures that resemble the angiosperms, demonstrating that there is a close evolutionary connection between the members of this group and the first angiosperms, which were called pro-angiosperms. Numerous examples indicate, with few exceptions, that this group was at the beginning unisexual and then became bisexual. The female strobilus was composed by eggs surrounded by sterile leaves and a tubular-shaped integument comprising the mycropile. The male reproductive structure was composed of leaves and structures containing small pollen sacs composed of fused sporangia (sinangio). The pollen grain, whose shape was monolete, resembled that of the cycads. In this group, the occurrence of both wind and self-pollination has been suggested, and some evidence indicated that animal pollination may also have been used (Willis and McElwain 2002).

In the Ginkgoales group the records of the first fossils date from 280 myr, and only a single species is currently found, *Ginkgo biloba*. Fossil evidence suggests that this group was composed of at least 16 genera and contributed significantly to global vegetation. The great similarity between extinct species and *Ginkgo biloba* has led to its description as a living fossil (Thomas and Spiecer 1987). The main eustele stem had a great content of secondary xylem and show characteristics that are difficult to separate from certain conifers such as *Pinus*. *Ginkgo* eggs are born at the end of short lateral branches, in number of two or three, and are connected by a peduncle. The microsporangiums (pollen sacs), however, are born in the leaf axils of short lateral branches (Mussa 2004). The reproductive strategy used by *Ginkgo biloba* is a dioecious type, but fossil evidence indicates that there was much variation between the reproductive structures (Willis and McElwain 2002).

Another important group was the Glossopteridaceae. This group has also been suggested as a possible ancestor of the angiosperms, as these plants, which had deciduous and arborescent habit, also presented highly modified reproductive organs attached to the leaves. Fossil evidence suggests that these plants grew up about 10 ft of height and had an eustele stem type, indicating close similarity with the modern *Araucaria*. Its female reproductive structures were quite diverse, ranging from uni- to multiovulate structures (Willis and McElwain 2002).

Conifers appeared in the Carboniferous period, and most of the primitive forms are currently extinct. The earliest conifer, found in fossil records in Yorkshire, *Swillingtonia denticulata*, was dated as belonging to the Upper Carboniferous (~310 myr). This group showed an increase in the fossil record during the Permian period, yet the largest radiation occurred in the Triassic period (245~208 myr), with seven families (Podocarpaceae, Taxaceae, Araucariaceae, Cupressaceae, Taxodiaceae,

Cephalotaxaceae and Pinaceae), which currently are widely dispersed. Among the conifers in evidence in the fossil records, *Utrechtia* is indicated as the most ancient plant. This plant, belonging to the Permian period, reached 5 m and possessed an eustele stem type with sap conducting vessels and tracheids. Its fossil specimen show morphological similarity to many existing conifers. The male and female reproductive structures of the plant were probably placed in different parts of the apex. Fossil evidence suggests that the pollen-producing male structure was very similar to current conifers (Thomas and Spiecer 1987). The female cones consisted of an axis of approximately 8 cm in length in which a reduced leave with the reproductive structure was found. This small fertile apex resembles an intermediate stage between the cordaites and the structure of modern conifers (Crane 1985; Doyle and Donoghue 1986). Thus, conifers, in general, can present a monoecious or dioecious reproductive strategy, containing male cones on the abaxial side of the leaf and egg cones in the bract surface (Mussa 2004).

1.5 Evolution of Angiosperms

As far as we know, the angiosperms diverged from a seed plant common ancestor between the late Jurassic and early Cretaceous between 130 and 90 myr (Crane et al. 1995), reaching dominance between 50 and 80 myr (Axelrod 1970), soon after the occurrence of the fifth mass extinction (Zunino and Zullini 2003). According to Stuessy (2004), the angiosperms originated from seed-producing fetuses (plants like ferns). These would have evolved in the Jurassic period, with the emergence of the carpel followed by the occurrence of double fertilization and only after these two evolutionary steps the mutations responsible for the appearance of the flower component parts (i.e., sepals and petals) should have occurred. This transition would have taken more than 100 myr to become complete.

The presence of extreme and inconstant weather would have restricted the location of these individuals to environments of higher altitudes and median latitudes, where there was a predominance of tropical dry weather until the early Cretaceous. However, with the fragmentation of Gondwana (~206 myr) the encroachment of the continents by ocean waters occurred (Lomolino et al. 2006), increasing the ocean surface around the continents. As a result, mild and homogeneous climate began for all continents, promoting the colonization of angiosperms in low altitude regions. On the other hand, the fragmentation and subsequent continental drift of Gondwana (originating South America, Africa, India and Australia) promoted the isolation of plants in the Cenomonian-Albian (Cretaceous intermediate phase) and significant climate change. The barriers to gene flow provided changes in allele frequencies and increased diversity due to pre-adaptive and adaptive functions (Axelrod 1970).

Morphological and molecular evidence and phylogeny studies have shown that the angiosperms had a monophyletic origin (Doyle and Donoghue 1986), with *Amborella trichopoda* being the most primitive angiosperm (*Nymphaeales*—aquatic lilies) and nearest sister group among them, from which the first diverging lineage would have appeared (Qiu et al. 1999; Soltis et al. 2000). On the other hand, two genera of Gnetales, *Gnetum* and *Welwitschia*, did not form a cluster with the angiosperms, showing a high level of consistency in the grouping with the conifers (Qiu et al. 2000). Fossil records confirm the possibility that all angiosperms have originated from a common ancestor derived from the gymnosperms and that this had no flowers, neither fused carpels, nor fruits. The oldest known fossil has been dated to 125 myr and was found by Dilcher et al. (2002), probably being the mother of all angiosperms. The fossilized material, belonging to the group of Liliaceae, has been found in China and was called *Archaefructus sinensis. Archaefructus* is considered a key fossil because it presents carpels; however, it does not have flowers. This absence of perianth parts and the presence of carpels and separate stamens, distributed along the axis in a vertical reproductive structure, posed questions about the possibility of the existence of unisexual flowers without perianth and more specialized forms in the base of the angiosperms (Friis et al. 2003). This information confirms the hypothesis that the angiosperms have started their evolution in the Lower Cretaceous (approximately 130 myr), reaching its dominance in the green world about 90 myr. Around 75 myr, many families existed already and some of the modern genera could also be found (Raven et al. 1995).

Currently, the angiosperms are worldwide dominant and have about 300–400 families and 240,000 to 300,000 thousand species, while the ferns have about 10,000 species and gymnosperms only about 750 species (Willis and McElwain 2002; Bernardes-de-Oliveira 2004). The appearance and rapid diversification of monocots and eudicots led these plants to an increasing dominance over the last 35 myr of the Upper Cretaceous (100~65 myr). At approximately 90 myr, several orders and families of existing flowering plants had already been established and the flowering plants had reached dominance throughout the Northern Hemisphere. In the following 10 myr they reached dominance in the Southern Hemisphere by having adaptive characteristics of drought and cold resistance such as smooth leaves of small size, presence of vessel elements with more efficient cells conducting sugar through the plant phloem and a resistant seed coat protecting the embryo against desiccation. The emergence of the deciduous habit also occurred early in the evolution of this group, allowing plants to become relatively inactive during periods of drought, extreme heat or cold, which probably contributed to the success recorded in the last 50 myr, when the climate in the world suffered frequent changes (Raven et al. 1995). One of the most important factors, for the angiosperms, has been the evolution of the reproductive system which allowed more precise pollination and a more specialized seed dispersal mechanism. Thus, individuals were able to occur widely dispersed in many different types of habitats such as desert, mountains and shallow waters, not developing only on open sea and polar regions. Other important factors of success were: a more developed autotrophic diploid phase, a reduced haploid stage, double fertilization and the development of carpels for a greater protection of the seeds (Paterniani 1974).

Thus, all angiosperms necessarily have flowers, fused carpels, double fertilization (responsible for the endosperm formation), microgametophytes with an extremely varied number of nuclei, stamens with two pairs of pollen sacs and the presence of sieve tubes and companion cells in the phloem (Friis et al. 1992; Bernardes-de-Oliveira 2004). So, the evolution of the angiosperms reported the presence of two new variants: the presence of floral whorls (sepals and petals), and the presence of both sexes in the same flower (hermaphroditism). Until this period (late Carboniferous—Early Cretaceous) flowers had no whorls, and each sex was located in a reproductive structure, on the same plant or separate plants. During evolution, the male and female strobiles, present in different locations and structures in gymnosperms, were found in a single arrangement. Additionally, the sepals and petals appeared, producing ornaments for the flowers which were beginning to evolve.

But how could this be possible?

Molecular studies have identified three factors (composed of one or more genes) that control the production of flower whorls, which were named factors A, B and C (Coen and Meyerowitz 1991). Currently, genes from gymnosperms that show a high similarity with floral initiation genes of flowering plants (transition from vegetative meristem to reproductive meristem) have also been isolated, suggesting the evolutionary conservation of the biological role of these genes (Lobo and Dornelas 2002). However, Kramer and Irish (2000) studying the level of conservation of these genes in lower eudicotyledons, magnoliids and monocotyledons, found that the ABC program is conserved only in some aspects while in others it showed a high degree of plasticity. Since genes are relatively well conserved in higher eudicotyledons, these may have been established only later in the evolution of angiosperms. In angiosperms these factors are responsible for the formation of sepals (Se), petals (Pe), stamens (St) and carpels (Ca) (Fig. 1.15), and in conifers (gymnosperms) its role is not well known yet. Apparently, factors B and C are the oldest existing in conifers. It is known that factor C alone determines the formation of carpels, but in association with factor B it is able to determine the differentiation of stamens. On the other hand, factor A alone determines the formation of sepals, while its association with factor B promotes the differentiation of the petals (Fonseca and Dornelas 2002). Thus, for the angiosperm flower to emerge it was required the presence of three factors (A, B and C) in association, and yet the evolution of a fourth factor, the transcription factor, called SUPERMAN (SUP), that should act on factor B to allow the expression of factor C to produce the carpels. Therefore, the evolution of angiosperms only became possible thanks to the presence of all these factors in combination (Fig. 1.15). For more details on this subject, see Chap. 2 of this book.

The primitive angiosperms presented solitary flowers at the end of the branches or loosely arranged in axes as in many *Paeonia* species, and the branches above the internodes had reduced leaves and secondary flowers (Fig. 1.16a). The developmental pattern of the floral axis and the formation of stamens and carpels differed greatly and showed a long period without differentiation of meristems, with the occurrence of only an increase in size and, subsequently, the differentiation into three regions: central initial zone, peripheral and apical dome of the meristem. Initially, the development of the perianth and androecium in the most primitive angiosperms was poorly differentiated in sepals and petals. The tapetum was probably composed of bracts and modified leaves. In a number of genera, such as: *Calycanthus*, *Paeonia*

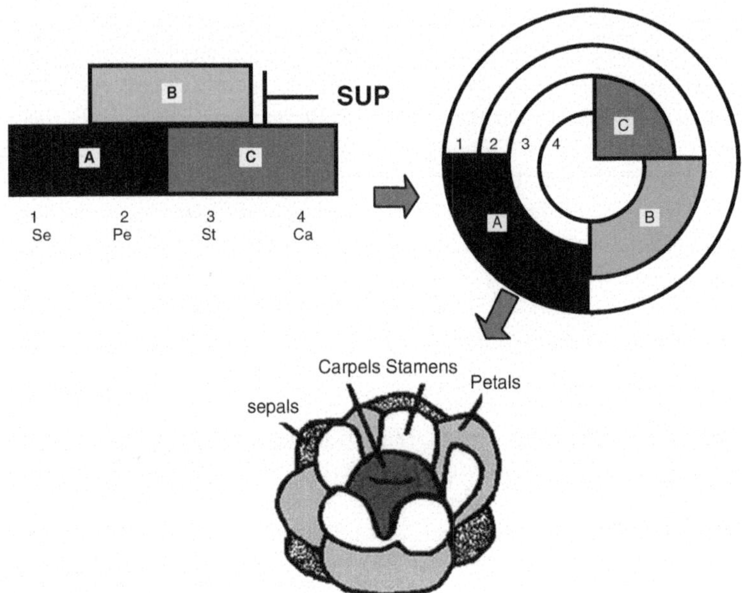

Fig. 1.15 Representation of the factors involved in the differentiation of vegetative meristems in the plant reproductive system (Karasawa et al. 2006)

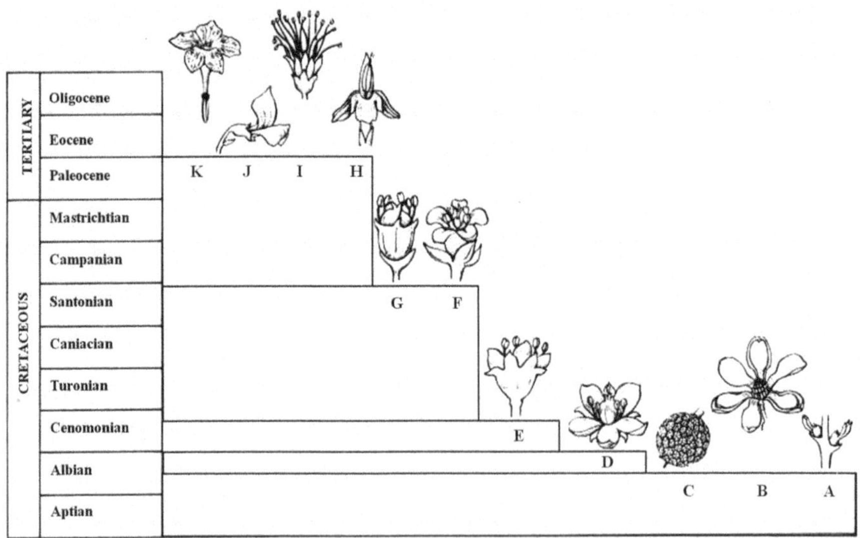

Fig. 1.16 Evolution of the flower types: (**a**) small with a few parts, (**b**) acyclic and hemicyclic, (**c**) monochlamydeous and unisexual, (**d**) cyclic, heterochlamydeous and actinomorphic, (**e**) epigynous and heterochlamydeous, (**f**) sympetalous, (**g**) epigynous and monochlamydeous, (**h**) zygomorphic, (**i**) form of a brush, (**j**) Papilionaceae family, (**k**) tube-shaped corolla (Friis et al. 1992, modified by Karasawa et al. 2006)

and some species of *Hibbertia*, gradual transitions occurred in the leaves, which changed from bract structures (modified leaves) to sepals and typical petals (Stebbins 1974). Studies of floral structures and reproductive organs from the Cretaceous and Tertiary also show a general increase in morphological and organizational diversity of the reproductive organs of angiosperms throughout evolution (Fig. 1.16a–f). However, the fossil records of floral organs are incomplete, especially in the early stages of diversification of this group, and this is also consistent with the records of other organs such as leaves and pollen (Friis et al. 1992).

The phyllotaxis of the floral parts of the Albian Stage (Lower Cretaceous) is unclear, but a few forms show evidence of a spiral arrangement of the parts. At the beginning of the Cenomanian (early Upper Cretaceous) the two major types of phyllotaxis in angiosperms were already established, comprising acyclic flowers with the parts arranged in spirals and hemicyclic flowers (Fig. 1.16b), with the parts of the perianth arranged in partially spiral whorls (Basinger and Dilcher 1984). From the beginning to the intermediate stage of the Cenomanian, acyclic and hemicyclic flowers were widely scattered among the angiosperms, and its importance declined with the diversification of cyclic flowers (Fig. 1.16d), which predominated in the fossil floras of the Santonian-Campanian stage (Upper Cretaceous), but the fossil pollen suggests that probably these forms were already established at the end of the Cenomanian stage.

The information on the number of floral parts in the reproductive structures of the Albian stage is also scarce, whereas the number of carpels varied from 3–8 to over 100. The known number of stamens is three and five, but unfortunately this is based on only two floral structures. Polymeric (with numerous parts), acyclic and hemicyclic flowers, with an indefinite number of parts, were apparently prevalent in the Cenomanian stage. On the other hand, cyclic flowers (Fig. 1.16d) presented, mostly, five parts, but some evidence suggests the existence of flowers with four and six parts also in this period. Apparently, the first cyclic flowers were isomerous (with the same number of floral parts in each whorl). The heteromerous, in turn, have only been established in an intermediate stage of the Cenomonian period and dominated the Santonian-Campanian stage having, usually, the perianth and androecium in a number of five and the gynoecium with two to three carpels. On the other hand, the true trimerous flower types were established and were relatively common in the Maastrichtian stage (late Upper Cretaceous). The perianth of the beginning of the Cenomanian was already established with different types of calyx and corolla. In relation to the symmetry and fusion of floral parts, it appears that all were apparently actinomorphic with radial symmetry and with the perianth parts free. The bilateral symmetry in flowers has occurred approximately 60 myr after the origin of angiosperms and is found in many fossil records of the Paleocene and Eocene and Upper Cretaceous when it was associated with the presence of social insects, and its co-evolution occurred in a number of families at different stages (Dilcher 2000). Fossils of zygomorphic flowers (Fig. 1.16h) were found only in the Maastrichtian stage, but evidence indicates that zygomorphism may have been established at the beginning of the Campanian stage. As to the distinct differentiation of the flower parts, it is known that this was found in late Paleocene in *Papilionoideae* flowers, while the first sympetalous flowers were observed in fossils of the Santonian-Campanian stage and a number of them were found in the Maastrichtian (belonging to the final stages of Cretaceous). In the Cretaceous, sympetalous flowers generally show a shallow and wide open tube form (Fig. 1.16f), while the deep tube forms were established in the Paleocene and early Eocene (Friis et al. 1992).

Based on the information originated from fossils, it can be inferred that flowers of the first angiosperm presented individual carpels, unisexual or bisexual small flowers, and radial symmetry (Friis et al. 1992). Also Dilcher (2000) has verified the presence of only small and medium-sized flowers, among the oldest fossil record of angiosperms. This author believes that the flower sizes had a relationship with the size of the pollinator, and the subsequent variation in the size of these flowers suggests a wide range of pollinators. The author states further that water and wind also participated in the process of pollination.

As for the position of the ovary, the structure of the fossils of angiosperm flowers and fruits of the Albian and early Cenomanian was a hypogeous type (Fig. 1.17a). The epigynous flowers (Fig. 1.17c) were found well established in the beginning and middle of the Cenomanian stage, and their radiation apparently reached its peak in the Santonian-Campanian stage, comprising around two-thirds of all the floral structures of this stage, decreasing in the Tertiary period. Currently, the epigeous flowers are present in one quarter of all families (Grant 1950).

With respect to the male reproductive structure, the first fossil records describe the existence of three stamens fused at the base, and also unisexual flowers with five stamens present in the Albian stage (intermediate stage of the Cretaceous) while in the Santonian-Campanian stage (Late Cretaceous) the stamens were well established in the fossil records. Fossil stamens with free filaments were found in the Cretaceous. The pollen sacs of all the anthers of the first known fossils have four sporangia, and only in the Santonian-Campanian stage the evidence of anthers with two sporangia appears. Dehiscence, initially, was longitudinal (Santonian-Campanian), while at the beginning of the Paleogene the first records of dehiscence by two or more valves were found, and in the Paleocene, the fossils presented the first anthers with apical dehiscence (Friis et al. 1992).

Pollen of the first angiosperms had a single opening, as found in monocots and in some other groups of angiosperms, as well as in the Cycadaceae, Ginkgoaceae

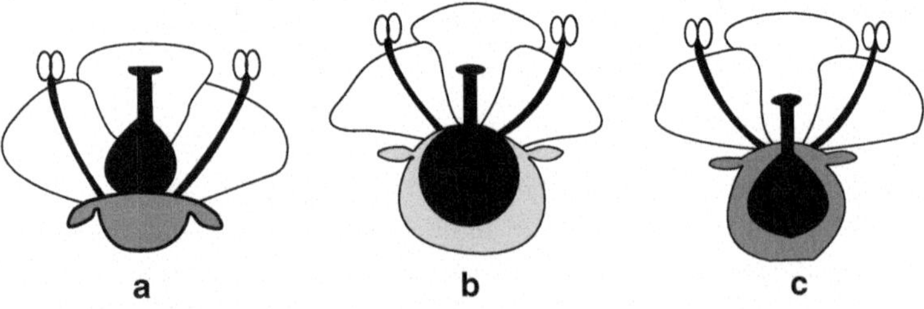

Fig. 1.17 Types of inflorescence according to the position of the ovary: (**a**) hypogeous; (**b**) perigynous; (**c**) epigeous (Karasawa et al. 2006)

and other groups. Currently, there are four known pollen types present in fossils found in the earliest angiosperms (*Clavatipollenites, Pre-Afropollis, Spinatus* and *Liliacidites*) and a fifth type (*Tricoliptes*) which can be found in the more recent angiosperms.

The female gametophyte has undergone major changes during the evolution of the angiosperms, with the evolution through modules (Fig. 1.18). It is believed that, in the beginning, the module was composed of four cells located in the micropyle region, comprising two synergids, one egg cell and one central cell, giving rise to diploid endosperm individuals. Williams and Friedman (2002) have shown that the presence of diploid endosperm was common in ancient lineages of angiosperms. Subsequently, the micropylar module would have suffered a duplication, which originated eight nuclei/seven cells, resulting in an embryo sac composed of the module located in the chalaza region (giving rise to three antipodes—which degenerate soon after fertilization), a central cell composed of two nuclei and the module located in the micropyle region (Friedman and Williams 2003). Thus, this is how individuals with triploid endosperm that presented 2:1 maternal/paternal ratio would have been originated (Williams and Friedman 2004). During evolution, the modules continued to be duplicated giving rise to endosperm with higher ploidy levels, with the presence of 1–14 nuclei in the polar region being sometimes detected (Friedman et al. 2008).

The fossil record of the female reproductive structure of the first angiosperms have shown free carpels of an apocarpic type, this being the prevailing condition at the beginning of the Albian and the Cenomanian (Fig. 1.19). The syncarpy (fusion of carpels), in turn, was only established at the end of the Albian (intermediate phase of the Cretaceous), and was represented by a number of taxa in the early Cenomanian. Basinger and Dilcher (1984) described a fossil of about 94 myr.

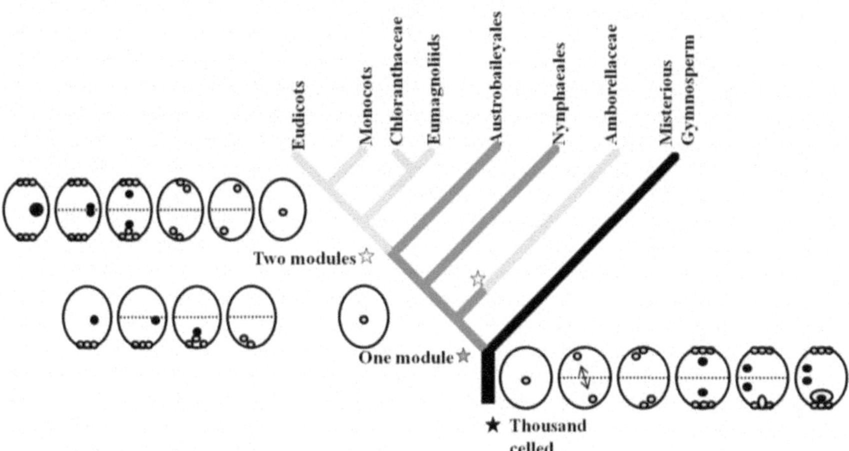

Fig. 1.18 Modular evolution of the female gametophyte (Drawn by Karasawa, MMG based on Williams and Friedman 2003, 2004)

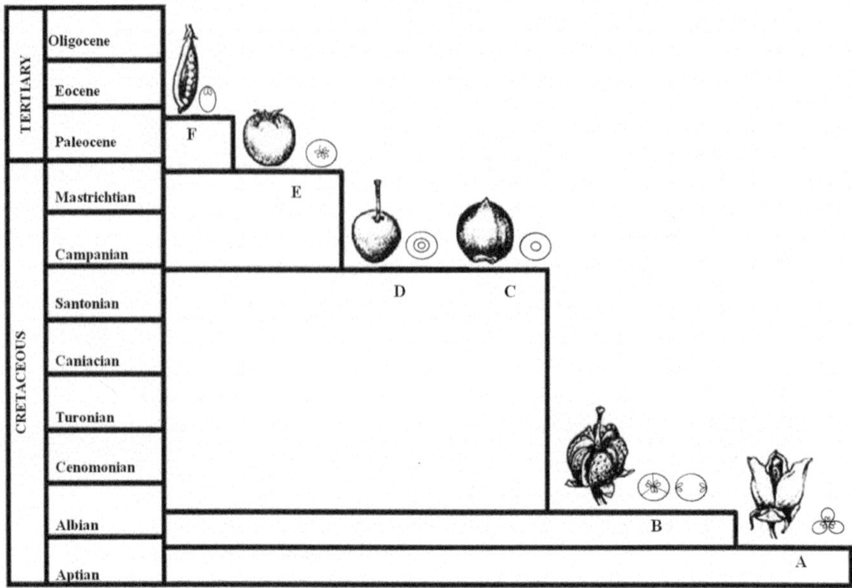

Fig. 1.19 Evolution of the fruit types: (**a**) follicles and nuts of apocarpic ovaries; (**b**) capsules; (**c**) nuts; (**d**) drupes; (**e**) berries; (**f**) pods. (**b–f**: syncarpic ovaries) (Friis et al. 1992, modified by Karasawa et al. 2006)

According to the authors, the fossil had pentamerous flowers with distinct sepals and petals, fused carpels and a floral receptacle. According to Friis et al. (1992), the syncarpic forms have become diversified at the end of the Cretaceous, being the most common reproductive structure in the Santonian-Campanian flowers.

The emergence of fused carpels was paramount in the evolution of angiosperms, this being the prevailing characteristic in their separation in relation to other seed plants. The fusion, almost always complete, has the task of protecting the unfertilized egg from the external environment. There are theories suggesting that the fusion of carpels occurred to promote protection against beetles and other herbivores. However, Dilcher (2000) suggests that this is more directly related to the evolution of the flowers bisexuality. With the flower evolution, the male and female organs were approximated, requiring, therefore, a protection against self-fertilization. To promote the necessary protection, mechanical (fusion of carpels) and chemical (self-incompatibility systems) barriers would have arisen, so that plants would prevent the growth of the pollen tube. Moreover, the addition or subtraction of sepals, petals and stamens was important to promote cross-pollination (outcrossing) and the emergence of nectaries was responsible for increasing the pollinization by insects.

Currently, 83 % of the taxa of existing angiosperms present syncarpy in the gynoecium (Endress 1982). Initially the syncarpic ovaries were, as it seems, partially separated and divided according to the number of loci corresponding to the number of carpels in the early Albian-Cenomanian stage. In the Santonian-Campanian, a number of different types with unilocular ovaries were developed, while the secondary divisions showed their first occurrence in the angiosperm fossils of the Maastrichtian stage (Friis et al. 1992). At this stage, "gynoecium" and apocarpic fruits (Fig. 1.19a) of the earliest known fossil record show no evidence of distinct "styles" and stigmatic area, as this type of characteristic shows inconsistency. The first syncarpic fruits (Fig. 1.19b–f) were found in the Albian-Cenomanian stage (lower third of the Cretaceous), where the fruits were apparently dry and without obvious modifications for dispersal. However, the evolution of syncarpy was relatively rapid and from the beginning to the end of the Cretaceous practically all kinds of syncarpic fruits were already established. The follicles and nuts were the fruits originated from apocarpic ovaries, while the separated capsules (Fig. 1.19b) were derived from syncarpic ovaries. The first fossils of fruits with pulp date from the middle third of the Cretaceous, while the first evidence of berries was found only in fossils belonging to the Maastrichtian stage (late Cretaceous) (Fig. 1.19e). The pulp fruits became relatively common during the early Paleogene period, increasing considerably their diversity in relation to size, indicating wide variation in dispersal mechanisms (Tiffney 1984).

It is believed that the first angiosperms experienced a variety of pollinators, being pollinated by water, wind or animals. However, it was the association with animals that provided their highest diversification (Bernardes-de-Oliveira 2004) throughout their evolutionary history.

1.5.1 Unisexuality and Reproductive Strategies

Today's angiosperms exhibit a great diversity of reproductive strategies. The vast majority of angiosperms present individuals with hermaphrodite flowers (72 %), while only 11 % of the plants have unisexual flowers (Fig. 1.20), 7 % being monoecious and 4 % dioecious (Figs. 1.21 and 1.22), whereas the intermediate forms of

Fig. 1.20 Types of flowers found in Angiosperms: female (*left*), hermaphrodite (*center*) and male (*right*) (Karasawa et al. 2006)

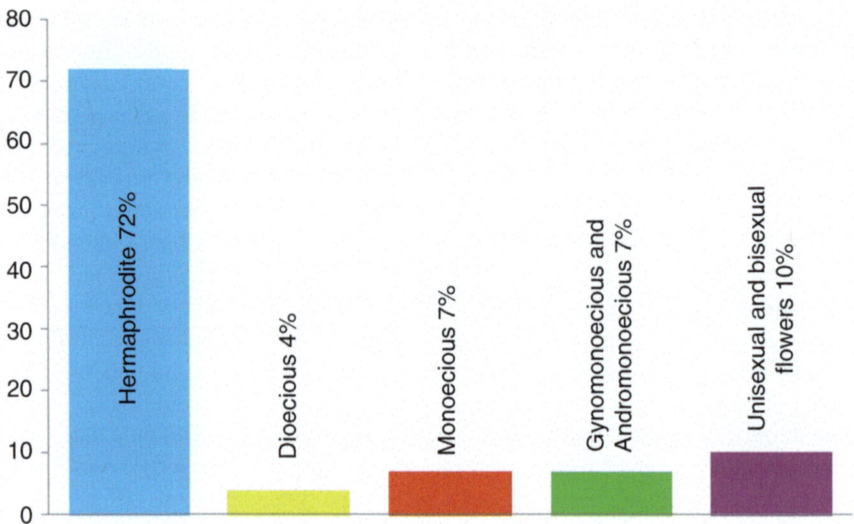

Fig. 1.21 Frequency of reproductive strategies present in the angiosperms (Karasawa 2005)

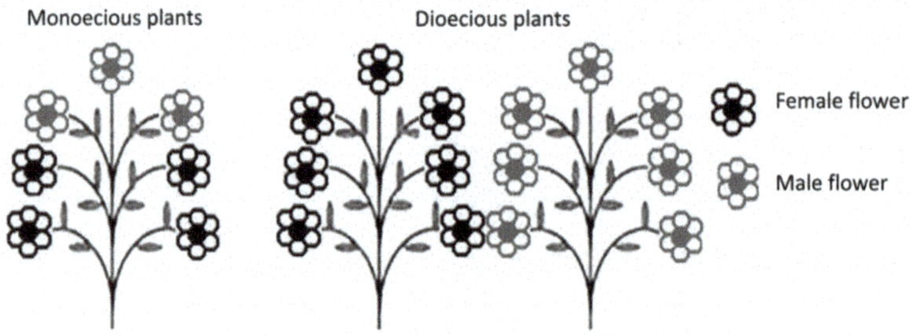

Fig. 1.22 Types of plants found in angiosperms: monoecious and dioecious (Karasawa et al. 2006)

sexual dimorphism (gynomonoecy and andromonoecy) represent 7 % and plants with both forms of unisexual and bisexual flowers comprise 10 % (Figs. 1.21 and 1.23) (Ainsworth 2000; Richards 1997).

The unisexual flowers can be found placed in different parts of a single plant (monoecious) or on different plants, forming dioecious populations (Fig. 1.22).

The monoecious populations may present plants with a gynomonoecious form (female and hermaphrodite flowers), andromonoecious (male and hermaphrodite flowers) or trimonoecious (male, female and hermaphrodite flowers). Likewise, dioecious populations may have gynodioecious forms (plants with female flowers and plants with hermaphrodite flowers), androdioecious (plants with male flowers and hermaphrodite plants), and also subdioecious forms (female flower plants, plants with male flowers and plants with hermaphrodite flowers) (Fig. 1.23).

But how could such a diversity of reproductive strategies in angiosperms have arisen?

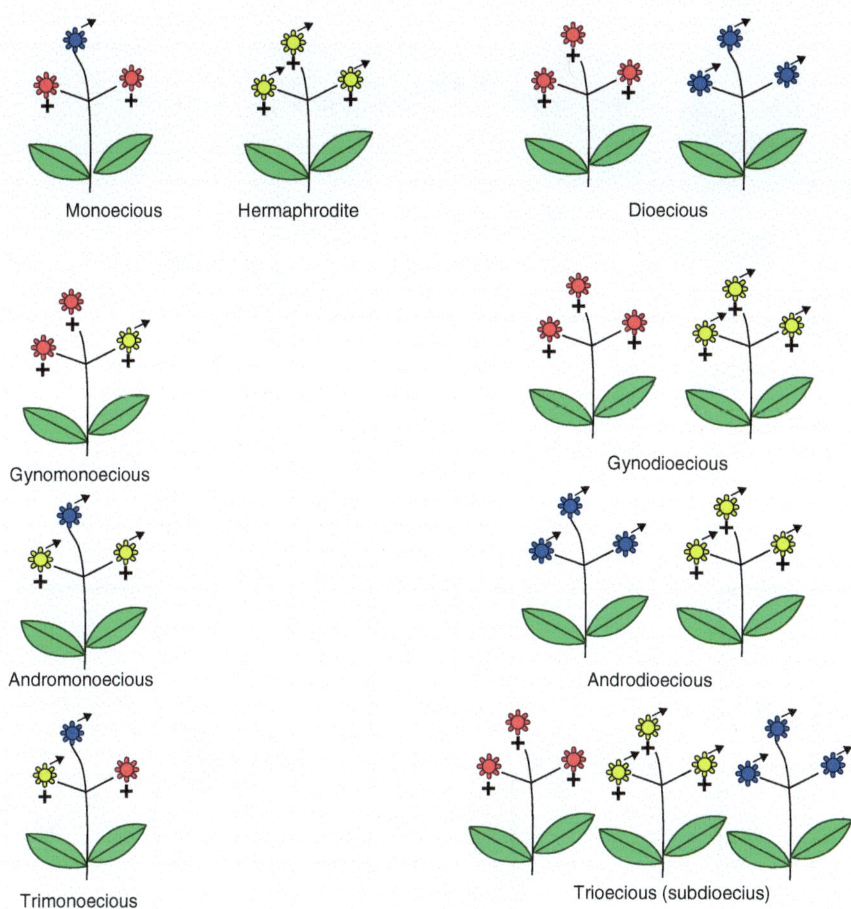

Fig. 1.23 Reproductive strategies adopted by angiosperms (Ainsworth 2000, modified by Karasawa 2005)

1.5.2 Evolution of Unisexuality

Unisexuality, in angiosperms, evolved as an outcrossing (cross fertilization) promoter system whose primary function is to obtain reproductive success in many different habitats in which they are found. The different evolutionary forces—selection, mutation, migration and drift—acting on hermaphrodite individuals throughout its evolution would have promoted the emergence of monoecious and dioecious populations (Barrett 2002).

- **Evolution of dioecy**

Dioecy, most of the times, evolved from self-compatible species (which can self-fertilize) in response to a selective pressure to promote outcrossing (Bawa and Opler 1975) and represents the change in the pattern of resource allocation for the

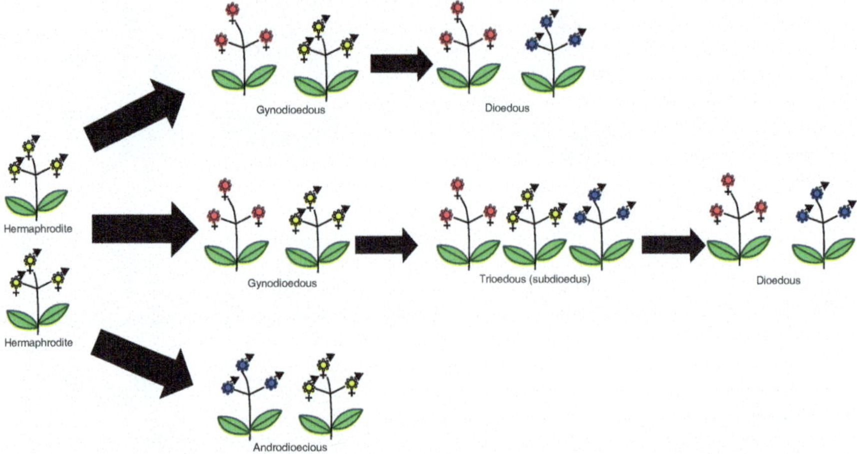

Fig. 1.24 Evolution of dioecy in hermaphrodite plants (Drawn by MMG Karasawa)

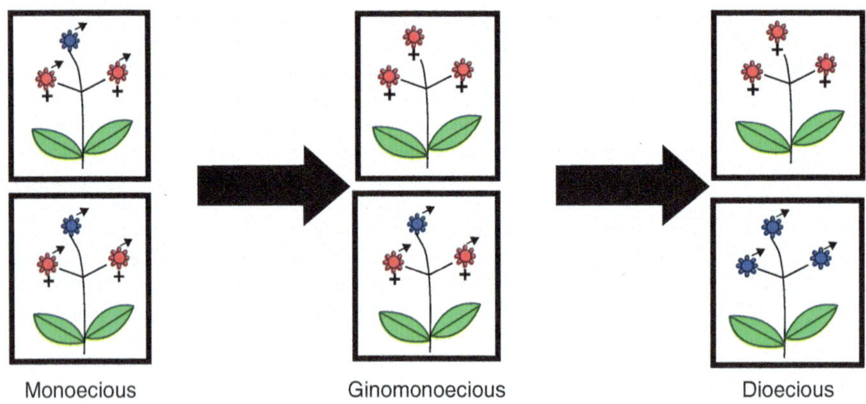

Fig. 1.25 Evolution of dioecy in monoecious plant populations (Drawn by MMG Karasawa)

male and female roles. Moreover, sexual dimorphism changes the spatial distribution of resources for pollinators, seed dispersers and predators (Bawa 1980; Sato 2002). Although Lebel-Hardenack and Grant (1997) believe that this evolution could have occurred only to allow a better allocation of resources optimizing reproduction. Ainsworth (2000) emphasizes that dioecy is one of the most extreme mechanisms that, most of the time, arises due to the deleterious effects of inbreeding depression or stressful environmental conditions and consequent resource limitations that prevent the hermaphrodite plants to maintain sexual functions and may promote the emergence of individuals with separate sexes. According to Charlesworth (1991), the evolution of dioecy may occur in hermaphrodite, monoecious and in populations that present heterostyly, as shown in Figs. 1.24, 1.25 and 1.26, respectively.

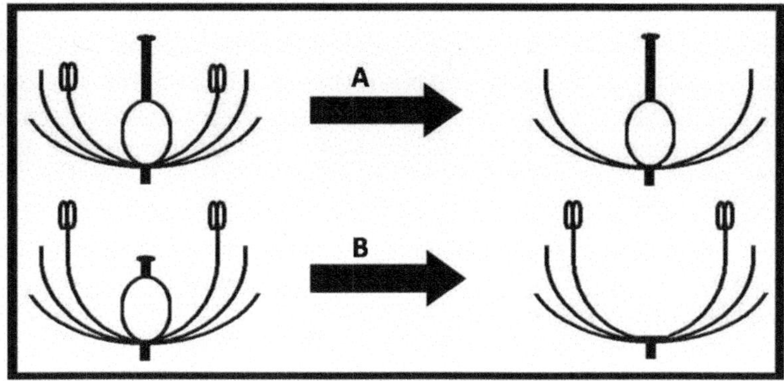

Fig. 1.26 Evolution of dioecy in plants with heterostyly (Karasawa et al. 2006)

In hermaphroditic populations, dioecy evolved as a result of at least two mutations, one causing male infertility which promotes the emergence of female plants, and another mutation causing female sterility making possible the appearance of male plants. Thus, one mutation would affect the production of pollen grains and the other the ovule production (Charlesworth and Charlesworth 1978, 1998; Charlesworth 1991). The authors believe it is unlikely that the two mutations would occur simultaneously for the establishment of dioecy, and that dioecy from hermaphroditism must have involved intermediate types in the population providing the presence of hermaphrodites, together with male and female sterile plants in the same population (subdioecious population) (Fig. 1.24). And that, in all cases in which the first mutation caused female sterility, with the emergence of androdioecious plants, the fall of dioecy would have been observed, because there is no registered case showing that this form have been able to evolve into established dioecious populations (Fig. 1.24). On the other hand, Sato (2002) reports based on mathematical models, have shown that those plants of separate sexes (dioecious) only become well established if there is a gradual reduction in male fertility or a reduction of the seeds production in hermaphrodite plants, providing the evolution of dioecy.

The evolution of dioecious plants, from monoecy, seems to involve only a single route, as male and female flowers already co-exist in a single plant, with the occurrence of mutations causing male and female sterility on different plants being sufficient. This would cause the separation of sexes in plants as shown in Fig. 1.25. Charlesworth and Charlesworth (1998) believe that for dioecy evolution to occur starting from monoecious populations, a series of mutations changing the proportions of male and female flowers in the plants are required until either sex is allocated on separate plants.

It is believed that distyly (a type of heterostyly) could originate dioecious plants due to the occurrence of mutations abolishing the male functions in some plants and female functions in others in order to give rise to plants of separate sexes (Lloyd 1979) (Fig. 1.26). One hypothesis is that the shift from distyly to dioecy is initiated by a change in pollination biology of these populations with the discontinuity of supplementary pollen between individuals which can occur in two ways: by promoting the flow between long stamens and pistils, and eliminating the usefulness and functionality of short stamens and pistils (Beach and Bawa 1980). However, it is worth highlighting that heterostyly may have had independent origins in plants pollinated by animals to increase the accuracy of pollination (Barrett et al. 2000).

- **Evolution of monoecy**

Monoecy and dioecy are quite different, as dioecy prevents selfing absolutely, while monoecy merely prevents pollination within the flower, but cannot prevent an individual to be self-fertilized.

Thus, just as plants have evolved as dioecious, monoecious plants may have originated from hermaphrodite plants by the suppression of male function in some flowers and the suppression of the female function in other flowers; however, this should have occurred in the same plant and have not been allocated to different plants as in the case of dioecy (Richards 1997).

Monoecious plants may also have been derived from dioecious plants, following the reverse path of dioecy (Fig. 1.25); however, this system must be considered with the onset of the female function in male plants, and vice versa for individuals in the same population, and in the end, these flowers had separate sexes in the same plant. However, it seems very unlikely that this evolutionary path has occurred.

The evolution of monoecious plants, from plants presenting heterostyly, seemed to be an easier and more likely mechanism, because this would involve the same steps discussed in the evolution of dioecy (Fig. 1.26); however, instead of unisexual flowers being placed in different individuals, these would be allocated in different parts of the same individual.

1.5.3 Evolution of Self-Incompatibility Systems

The origin and maintenance of the self-incompatibility systems are quite complex and there are still many questions surrounding its evolution that remain unanswered. It is believed that it arose several times during evolution, and to assure the establishment of the chemical system of self-incompatibility, the occurrence of strong inbreeding depression is necessary in individuals from self-compatible populations. Moreover, plants completely self-incompatible would not be established immediately in the population of self-compatible plants. It is believed that initially intermediate levels in the population would have been established, with compatible plants being among self-incompatible plants. Because the generated individuals did not show inbreeding depression, the self-incompatibility alleles would have a reproductive advantage by increasing its frequency in the population as to completely suppress the self-compatibility alleles. Another fact to consider is that alleles responsible for self-incompatibility do not increase in frequency if there is no inbreeding depression, because in the absence of depression both types of alleles would have the same reproductive advantage and, therefore, the individuals generated from self-compatible plants would not be eliminated (Clark and Kao 1994).

Thus, the reproductive advantage can be defined by the balance of two forces: one that controls the rejection of the pollen grain, prioritizing outcrossing, and that which acts in the opposite direction as to increase the frequency of progenies in environments where the presence of pollinators is low. So, the maintenance of self-incompatibility throughout generations will depend on the superiority of the progenies produced and the cost for reducing the number of individuals generated (Vallejo-Marín and Uyenoyama 2004).

1.5.4 Evolution of the Self-Fertilization System

The breakdown of self-incompatibility has occurred repeatedly during the evolution of angiosperms and provoked profound impacts on the genetic structure of populations (Stone 2002). The primary genetic cost of inbreeding is the deleterious effect of depression; however, it is not constant and varies depending on the level of self-fertilization. Allogamous populations that practice self-fertilization suffer the effect of depression generated by exposure of deleterious or partly deleterious recessive alleles until they are completely eliminated from the population. From this moment on, the population is prepared to continue to evolve, using the system of self-fertilization without having recurring detrimental effects.

Considering a self-incompatibility system associated with the presence of an ancestral clonality, it was found that the transition from self-incompatible clonal reproduction (SI C) to self-incompatible non-clonal reproduction (SI NC) rarely occurs, although the reverse is common (Fig. 1.27). However, the transition from SI

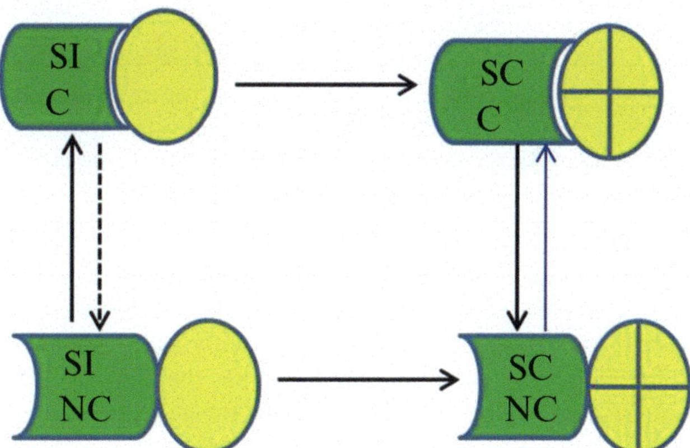

Fig. 1.27 Evolution of the self-incompatibility system from the self-compatible ancestral (Drawn by MMG Karasawa based on Vallejo-Marín and O'Brien 2007)

C to self-compatible clonal reproduction (SC C) commonly occurs and is irreversible. Similarly, the SI NC system undergoes transition to autocompatible non-clonal reproduction (SC NC) is irreversibly. In contrast, the transition between SC C to SC NC occurs frequently, while the reverse varies according to environmental conditions (Vallejo-Marín and O'Brien 2007).

1.5.5 Evolution of the Mixed Mating System

The mixed mating system is common in higher plants (Ingvarsson 2002) and corresponds to the simultaneous occurrence of self-fertilization and cross-fertilization. Currently, there is strong evidence that it is generated mainly by high inbreeding depression (Goodwillie et al. 2005). The primary genetic cost of inbreeding is the deleterious effect of depression, however it is not constant and varies depending on the level of self-fertilization. In a condition where the inbreeding depression condition does not vary and its fluctuation occurs in a stochastic way between generations, with an average of approximately 0.5, self-fertilization is not necessarily selected. As a result of this fluctuation, inbreeding depression can be seen as an additional cost of self-fertilization that can be stabilized in a mixed mating system (Cheptou and Schoen 2002). The substantial frequency of species with intermediate crossing rates provides evidence that the mixed mating system can be a stable strategy; yet there is no measurement of this frequency, so that it cannot be stated that this mode of reproduction is stable or just a transition phase. For this, theoretical studies are needed with a larger number of taxa for greater precision in the conclusions (Goodwillie et al. 2005).

1.6 Evolutionary Implications

Along its evolution, plants have adapted to different forms of sexual and asexual reproduction. The knowledge of the mode of reproduction of the species is important because it has great effect on the colonization of different habitats and also in response to environmental changes.

To understand the evolutionary significance of the different mating systems, consider that in a population three different mutations arise that do not affect fertility or survival of the species, but the association of these may provide evolutionary advantage in the adaptation of the species. Assuming initially that the population presents asexual reproduction (every individual produced has the genotype identical to the mother), they would have much difficulty to gather the different mutations if they occur in different individuals, unless individuals already possessing a mutation would acquire new mutations. Therefore, it would require a long time until all individuals in the population would present all three mutations. Moreover, populations of sexual reproduction would quickly gather the different mutations that occur in different individuals by cross-fertilization, through the exchange of alleles between individuals within a short time, thus benefiting from faster selective advantage. On the other hand, we must consider that the population size also plays a fundamental role in the dynamics of dispersal of each new mutation, because if populations are too small mutations are not likely to be fixed and dispersed among the individuals of the population, and are usually lost by drift on both sexual and asexual forms (Crow and Kimura 1965; Hartl and Clark 1997).

The impact of the content and distribution of genetic variation within and among populations may represent an important role in the distribution of different characters, determining the extent and pattern of responses to natural selection. The reproductive system has shown that it plays a prominent role in this respect. Autogamous populations, due to the high level of homozygosity, have no potential variability within populations, because the mutations that arise are eliminated faster than in allogamous populations, which may limit their ability to respond to environmental changes. In general, it is expected that autogamous and asexual species have short lifespan (Holsinger 2000), because the offspring may not survive to reproduce, thus producing discontinuity in seed production (Herlihy and Eckert 2002). The size of the population also has an important effect on the diversity and on genetic risks of extinction, because small populations are more likely to lose alleles important for adaptation by drift, and more likely to cross between related individuals suffering the deleterious effects of inbreeding, increasing the risk of extinction as the levels of inbreeding increase (Blisma et al. 2000).

Moreover, it is important to consider that in nature rarely pure breeding systems are found (i.e., plants with a single reproductive system). A study evaluating the correlated evolution of self-incompatibility and clonal reproduction in Solanaceae performed by Vallejo-Marín and O'Brien (2007) found that there is a strong association between these forms of reproduction, and that all self-incompatible species of *Solanum* present clonal reproduction, supporting the hypothesis that clonal reproduction promotes reproductive success in the evolution of reproductive strategies in plants. According to the results clonality leads to benefits in colonizing species such as the Solanaceae, supporting the persistence of self-incompatible genotypes in the case of inhospitable environments where the presence of the pollinator is rare and/or the level of incompatibility is high reducing the sexual outcrossing progeny produced. On the other hand, clonality generates the aggregation of similar genotypes and this can lead to breakage of self-incompatibility system throughout evolution. The resolution of this paradox between clonality × evolution of self-incompatibility (Fig. 1.28) lies in the degree to which clonal propagation would be compensating or limiting seed reproduction, and the extent to which clonality would reduce pollen flow between established genotypes, which would be affected by clonal architecture, plant density and the type and presence of pollinator (Vallejo-Marín 2007).

The effects of the pollinating agent have also proven to be effective in changing the distribution of sexual reproductive systems in nature. It has been found that anemophilous species (i.e., wind pollinated species) present a bimodal distribution, that is, either autogamous or allogamous modes of reproduction, with little or

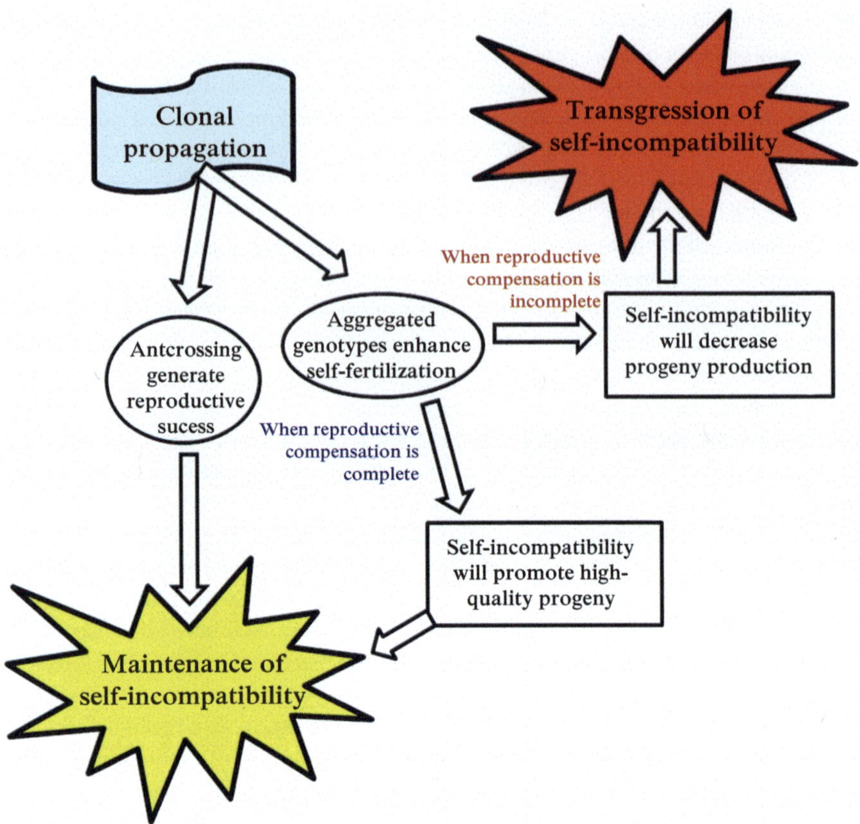

Fig. 1.28 Paradox: clonality × self-incompatibility (Drawn by MMG Karasawa based on Vallejo-Marín 2007)

rare intermediate types. Moreover, animals pollinated species present a continuous distribution between the two types of reproductive system, ranging from autogamous forms to extreme allogamous forms, with all degrees of self-fertilization and outcrossing in the middle (Vogler and Kalisz 2001).

Finally, it has been found that modular evolution of cells present in the female gametophyte has led to an increased ploidy in the produced endosperm. And that this mechanism is evolutionarily beneficial and stable, presenting the following consequences: increased level of heterozygosity in the endosperm, reduced genomic conflict (increase in the maternal/paternal relation) and increase in the range of observed phenotypes. This increase in ploidy level leads to a higher heterozygosity level that, in turn, would have an effect on the nutrition and vigor of the formed embryo. Furthermore, embryos have been more vigorous when the polyploid endosperm was generated from allogamous crosses among unrelated individuals. Thus, it is believed that selection should favor the evolution of individuals presenting endosperm containing higher ploidy levels (Friedman et al. 2008).

Bibliography

Ainsworth C (2000) Boys and girls come out to play: the molecular biology of dioecious plants. Ann Bot 86:211–221

Andrews HN (1963) Early seed plants. Science 142:925–931

Axelrod DI (1970) Mesozoic paleogeography and early angiosperm history. Bot Rev 36:277–319

Banks HP (1970) Evolution of the plants on the past. Wadsworth Pub. Co., Belmont, 170 p

Barrett SCH (2002) The evolution of plant sexual diversity. Nature 3:274–284

Barrett SCH, Jesson LK, Baker AM (2000) The evolution and function of stylar polymorphisms in flowering plants. Ann Bot 85(Suppl A):253–265

Basinger JF, Dilcher DL (1984) Ancient bisexual flowers. Science 224:511–513

Bateman RM, Crane PR, DiMichele WA, Kenkrick PR, Rove NP, Speck T, Stein WE (1998) Early evolution of land plants: phylogeny, physiology and ecology of the primary terrestrial radiation. Annu Rev Ecol Syst 29:263–292

Bawa KS (1980) Evolution of dioecy in flowering plants. Annu Rev Ecol 11:15–39

Bawa KS, Opler PA (1975) Dioecism in tropical forest trees. Evolution 29:167–179

Beach JH, Bawa KS (1980) Role of pollinators in the evolution of dioecy from distyly. Evolution 34(6):1138–1142

Bernardes-de-Oliveira MEC (2004) A origem e a evolução das angiospermas. In: Carvalho I de S (ed) Paleontologia, vol 1. 2nd edn. Interciência, Rio de Janeiro, pp 509–542

Blisma R, Bundgaard J, Boerema AC (2000) Does inbreeding affect the extinction risk of small populations? predictions from *Drosophila*. J Evol Biol 13:502–514

Brown JH, Lomolino MV (2005) Biogeografia, 2nd edn. Funpec, Ribeirão Preto, 691 p

Charlesworth B (1991) The evolution of sex chromosomes. Science 25:11030–11033

Charlesworth B, Charlesworth D (1978) A model for evolution of dioecy and gynodioecy. Am Nat 112:975–997

Charlesworth B, Charlesworth D (1998) Some evolutionary consequences of deleterious mutations. Genetica 102(103):3–19

Cheptou PO, Schoen DJ (2002) Frequency-dependent inbreeding depression in Amsinckia. Am Nat 162(6):744–753

Clark AG, Kao T-H (1994) Self-incompatibility: theoretical concepts and evolution. In: Clarke AE, Knox BR, Williams EG (eds) Genetic control of self-incompatibility and reproductive development in flowering plants. Kluwer Academic, Ordrecht, pp 220–244

Coen ES, Meyerowitz EM (1991) The war of verticilos: genetic interactions controlling flower development. Nature 353:31–37

Crane PR (1985) Phylogenetic analysis of seed plants and the origins of angiosperms. Ann Mo Bot Gard 72:716–793

Crane PR, Friis EM, Pedersen K (1995) The origin and early diversification of angiosperms. Nature 734:27–33

Crow JF, Kimura M (1965) Evolution in sexual and asexual populations. Am Nat 99:439–450

Dilcher D (2000) Toward a new synthesis: major evolutionary trends in the angiosperm fossil record. Proc Natl Acad Sci U S A 97(13):7020–7036

Dilcher D et al (2002) Oldest flower found in China. http://news.bbc.co.uk/1/hi/world/asia-pacific/1966248.stm

Doyle JA, Donoghue MJ (1986) Seed Plant phylogeny and the origins of angiosperms: an experimental cladistic approach. Bot Rev 52:321–431

Drews GN, Ydegari R (2002) Development and function of the angiosperm female gametophyte. Annu Rev Genet 36:99–124

Edwards DS (1986) Agalophyton major, a non-vascular land-plant from the Devonian Rhynie Chert. Bot J Linn Soc 93:19–36

Edwards D, Banks HP, Ciurca JR, Laub RS (2004) New Silurian Cooksonias from dolostomes of north-eastern North America. Bot J Linn Soc 146:399–413

Endress PK (1982) Sincarpy and alternative modes of escaping disadvantages of apocarpy in primitive angiosperms. Taxon 31:48–52

Fonseca TC, Dornelas MC (2002) Evolução do sexo em plantas. Biotecnol 27:48–51

Freeman S, Herron JC (1998) Evolutionary analysis, 2nd edn. Prentice Hall, Upper Saddle River, 702 p

Friedman WE, Williams JH (2003) Modularity of the angiosperm female gametophyte and its bearing on the early evolution of endosperm in flowering plants. Evolution 57(2):216–230

Friedman WE, Madrid EN, Williams JH (2008) Origin of the fittest and survival of the fittest: relating female gametophyte development to endosperm genetics. Int J Plant Sci 169(1):79–92

Friis EM, Chaloner WG, Crane PR (1992) The origins of angiosperms and their biological consequences. Cambridge University, New York, 358 p

Friis EM, Doyle JA, Endress PK, Leng Q (2003) *Archeafructus*—angiosperm precursor of specialized early angiosperm. Trends Plant Sci 8:369–373

Gensel PG, Andrews HN (1987) The early land plants. Am Sci 75:478–489

Goodwillie C, Kalisz S, Eckert CG (2005) The evolutionary enigma of mixed mating systems in plants: occurrence, theoretical explanations, and empirical evidence. Annu Rev Ecol Evol Syst 36:47–79

Graham LE, Cook ME, Busse JS (2000) The origin of plants: body plan changes contributing to a major evolutionary radiation. Proc Natl Acad Sci U S A 97(9):4535–4540

Grant V (1950) The protection of ovules in flowering plants. Evolution 4:179–201

Gray J, Shear W (1992) Early life on land. Am Sci 80:444–456

Hartl DL, Clark AG (1997) Principles of population genetics, 3rd edn. Sinauer Associates, Sunderland, no. páginas 519 p

Herlihy CR, Eckert CG (2002) Genetic cost of reproductive assurance in a self-fertilizing plant. Nature 416:320–323

Holsinger KE (2000) Reproductive systems and evolution in vascular plants. Proc Natl Acad Sci U S A 97(13):7037–7042

Ingvarsson PK (2002) A metapopulation perspective on genetic diversity and differentiation in partially self-fertilizing plants. Evolution 56(12):2368–2373

Karasawa MMG (2005) Análise da estrutura genética de populações e sistema reprodutivo de *Oryza glumaepatula* por meio de microssatélites. Tese (Doutorado)—Escola Superior de Agricultura "Luiz de Queiroz", Universidade de São Paulo, Piracicaba, 91 p

Karasawa MMG, Oliveira GCX, Veasey EA, Vencovsky R (2006) Evolução das plantas e de sua forma de reprodução. M.M.G.Karasawa, Piracicaba, 86 p

Kenrick P, Crane PR (1997) The origin and early evolution of land plants. Nature 389:33–39

Knoll AL (1992) The early evolution of eukaryotes: a geological perspective. Science 256:622–627

Kramer EM, Irish VF (2000) Evolution of the petal and stamen developmental programs: evidence from comparative studies of the lower eudicots and basal angiosperms. Int J Plant Sci 161(Suppl 6):S29–S40

Lebel-Hardenack KS, Grant SR (1997) Genetics of sex determination in flowering plants. Trends Plant Sci 2:130–136

Lloyd DG (1979) Evolution towards dioecy in heterostylous plants. Plant Syst Evol 131:71–80

Lobo CA, Dornelas MC (2002) Biologia molecular do desenvolvimento reprodutivo em *Pinus*. Biotecnol 28(9/10):40–43

Lomolino MV, Riddle BR, Brown JH (2006) Biogeography, 3rd edn. Sinauer Associates, Sunderland, 845 p

McAlester AL (1978) A história geológica da vida, 3rd edn. Edgard Blücher, São Paulo, 173 p

Mussa D (2004) Paleobotânica: conceituação geral e grupos fósseis. In: Carvalho I de S Paleontologia, vol 1. 2nd edn. Interciência, Rio de Janeiro, pp 413–508

Paterniani E (1974) Evolução dos sistemas reprodutivos. In: Salzano FA (ed) Natureza do processo evolutivo, vol 26, no. 5. Ciência e Cultura, pp 476–481

Qiu Y-L, Lee J, Bernasconi-Quadroni F, Soltis DE, Soltis PS, Zanis M, Zimmer EA, Chen Z, Savolainen V, Chase MW (1999) The earliest angiosperms. Nature 402:404–407

Qiu Y-L, Lee J, Bernasconi-Quadroni F, Soltis DE, Soltis PS, Zanis M, Zimmer EA, Chen Z, Savolainen V, Chase MW (2000) Phylogeny of basal angiosperms: analyses of five genes from three genomes. Int J Plant Sci 161(Suppl 6):S3–S27

Raven PH, Evert RF, Eichorn SE (1995) Biologia vegetal, 5th edn. Guanabara Koogan, Rio de Janeiro, 728 p

Raven PH, Evert RF, Eichorn SE (2007) Biologia vegetal, 7th edn. Guanabarra Koogan, Rio de Janeiro, 830 p

Richards AJ (1997) Plant breeding systems, 2nd edn. Chapman & Hall, Cambridge, 529 p

Sato H (2002) Invasion of unisexuals in hermaphrodite populations of animal-pollinated plants: effects of pollination ecology and floral size-number trade-offs. Evolution 56(12):2374–2382

Schopf JM (1968) Microfossils of the early Archean apex chert: new evidence of antiquity of life. Science 260:640–650

Schopf JM, Packer BM (1987) Early Archean (3.3 billion to 3.5 billion-year-old) microfossils from Warrawoona Group, Australia. Science 237:70–73

Soltis PS, Soltis DE, Zanis MJ, Kim S (2000) Basal lineages of angiosperms: relationships and implications for floral evolution. Int J Plant Sci 161(Suppl 6):S97–S107

Stebbins GL (1974) Flowering plants: evolution above the species level. Belknap, Cambridge, 399 p

Stone JL (2002) Molecular mechanisms underlying the breakdown of gametophytic self-incompatibility. Q Rev Biol 77(1):17–32

Stuessy TF (2004) A transitional-combinational theory for the origin of angiosperms. Taxon 53(1):3–16

Thomas BA, Spiecer RA (1987) The evolution and palaeobiology of land plants, 1st edn. Croom Helm, London, 309 p

Tiffney BH (1984) Seed size, dispersal syndromes, and the rise of angiosperms: evidence and hypothesis. Ann Mo Bot Gard 71:551–576

Vallejo-Marín M (2007) The paradox of clonality and the evolution of self-incompatibility. Plant Signal Behav 2(4):e1–e2

Vallejo-Marín M, O'Brien HE (2007) Correlated evolution of self-incompatibility and clonal reproduction in *Solanum* (Solanaceae). New Phytol 173:415–421

Vallejo-Marín M, Uyenoyama MK (2004) On the evolutionary costs of self-incompatibility: Incomplete reproductive compensation due to pollen limitation. Evolution 58:1924–1935

Vogler DW, Kalisz S (2001) Sex among the flowers: the distribution of plant mating systems. Evolution 55(1):202–204

Williams JH, Friedman WE (2002) Identification of diploid endosperm in an early angiosperm lineage. Nature 415:522–526

Williams, JH; Friedman, WE. 2003. Modularity of the angiosperm female gametophyte and its bearing on its early evolution of endosperm in flowering plants. Evolution, 57(2):216-230

Williams JH, Friedman WE (2004) The four-celled female gametophyte of *Illicium* (Illiciaceae; Austrobaileyales): implications for understanding the origin and early evolution of monocots, eumagnoliids, and eudicots. Am J Bot 91(3):332–351

Willis KJ, McElwain JC (2002) The evolution of plants. Oxford University, New York, 378 p

Zunino M, Zullini A (2003) Biogeografía: la dimensión espacial de la evolución. Fondo de Cultura Mexicana, México, 358 p

Chapter 2
Biology and Genetics of Reproductive Systems

Marines Marli Gniech Karasawa, Marcelo Carnier Dornelas, Ana Cláudia Guerra de Araújo, and Giancarlo Conde Xavier Oliveira

Abstract Chapter 2 will discuss the biology and genetics of asexual and sexual plant reproduction. In relation to the asexual system, special attention will be given to the apomictic reproduction, where we will point out aspects of this system, as well as details of the knowledge about mechanisms and genes involvement. Regarding the sexual reproduction system, we will discuss details of this reproductive mode, the life cycle of plants, the gametophytic and sporophytic generation, as additionally to the control of genes in reproductive organs. Subsequently, aspects of the mechanisms that promote allogamy, selfing and mixed mating will be presented.

2.1 Introduction

The reproductive system of angiosperms, in general, can be classified as sexual and asexual. The asexual forms include all the mechanisms that originate clones that are genetically identical to the mother plant. Moreover, sexual forms have different cross systems, which are: autogamous systems (i.e. plants with self-fertilization), allogamous (plants which present cross-fertilization, usually self-incompatible) and mixed (plants that present self-fertilization and cross-fertilization) (Fryxel 1957). The frequency of sexual and asexual reproductive systems is summarized in Fig. 2.1.

M.M. Gniech Karasawa (✉) • G.C.X. Oliveira
Department of Genetics, "Luiz de Queiroz" College of Agriculture, University of Sao Paulo, Avenida Pádua dias, 11, Piracicaba, SP, 13418-900, Brazil
e-mail: mgniechk@gmail.com

M.C. Dornelas
Department of Plant Biology, Biology Institute, State University of Campinas, Rua Monteiro Lobato 255, Campinas, SP, 13083-862, Brazil

A.C.G. de Araújo
EMBRAPA—Recursos Genéticos e Biotecnologia, Predio de Biotecnologia, Parque Estação Biológica—PqEB, Av. W5 Norte (Final) Caixa Postal 02372, Brasília, DF, 70770-917, Brazil

M.M. Gniech Karasawa (ed.), *Reproductive Diversity of Plants*,
DOI 10.1007/978-3-319-21254-8_2

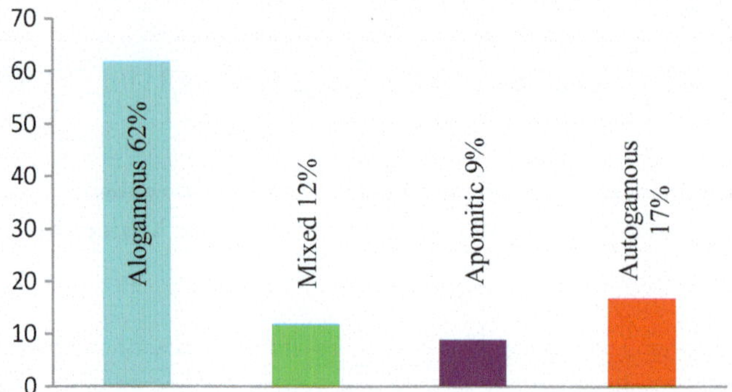

Fig. 2.1 Frequency of sexual and asexual reproductive systems (Karasawa 2005)

2.2 Asexual Reproduction

The asexual plant reproduction comprises all types in which there is either no involvement of gametes or only partial participation that results in a progeny identical to only one parent or entirely derived from only one parent. This occurs by mitosis of somatic cells or the egg cell but also by abnormal forms of meiosis and can be classified into two broad categories: vegetative reproduction and apomictic reproduction (Cavalli 2003; Raven et al. 2007).

2.2.1 Vegetative Reproduction

Vegetative reproduction is one of the asexual reproduction forms commonly found in plants. This type of reproduction consists in the production of new individuals originated from different parts of maternal tissues, without the involvement of any reproductive organ (Cavalli 2003). There are several strategies used by plants to promote this type of reproduction, namely: runners or stolons, rhizomes or underground stems, corms, bulbs or tubers, roots or shoots and leaves (Hartmann and Kester 1975; Raven et al. 2007).

Runners or Stolons

Runners comprise stems growing horizontally at the soil surface or just below ground. They are also known as stolons and originate new plants that are genetically identical to the maternal parent (Raven et al. 2007). This type of asexual reproduction is present in strawberry (*Fragaria* sp.) and violet (*Viola* sp.), for example.

Rhizomes and Underground Stems

Rhizomes are underground stems which grow and develop more or less parallel to the soil surface. Besides the adventitious roots, this type of stem also produces shoots that will give rise to new plants genetically identical to the maternal parent. They are important vegetative organs in the reproduction of species such as ferns and orchids. Moreover, they can play a role of storage in some plants (Raven et al. 2007), such as in sugar cane (*Saccharum* spp.) (Hartmann and Kester 1975).

Corms, Bulbs or Tubers

Corms, bulbs and tubers are organs of vegetative reproduction known as specialized storage stems. They have meristematic structures called buds that can give rise to new shoots and, consequently, to new plants that are genetic clones of the maternal parent (Raven et al. 2007). An example of plant with corms is the gladiolus (*Gladiolus* sp.); bulbs are found in lilies (*Lilium* sp.) and *Tulipa* sp.; and tubers in potatoes (*Solanum tuberosum*) and dahlia (*Dahlia* sp.) (Hartmann and Kester 1975).

Roots or Shoots

Roots and shoots are underground stems produced from roots of certain plants giving origin to new plants. Moreover, they can be originated as upright stems from the basis of branches. They are popularly known as "bud hickey" or "thief" (Raven et al. 2007). These reproduction forms can be observed in raspberry (*Rubus idaeus*), cherry (*Prunus* sp.), blackberry (*Rubus* sp.), apple (*Malus domestica*) and banana (*Musa* sp.).

Leaves

In some species, leaves may also play the reproductive function. This type of reproduction is frequently found in many individuals of *Kalanchoë daigremontiana* that produces numerous seedlings from the meristematic tissue located on the boundary of the leaves and in the fern *Asplenium rhizophyllum*, which originates new plants by means of leaf rooting. Whenever these roots reach a certain stage of development, the plantlets are dropped in the soil to start new roots (Raven et al. 2007).

2.2.2 Apomictic Reproduction

The apomictic processes are described in more than 400 genera belonging to 40 families, being prevalent in Poaceae, Asteraceae, Rosaceae and Rutaceae (Bashaw 1980; Hanna and Bashaw 1987; Carman 1997). It can be defined as the pathway that produces viable seeds without the fusion of gametes—agamospermy or seed production in the absence of sex (Hartmann and Kester 1975; Brown and Emery 1958; Nogler 1984; Appels et al. 1998). An important difference between the apomictic and sexual mode of reproduction is that the embryo formed by the latter is the result of the recombination of the male and female gametes, while the apomictic is derived only from maternal tissues, lacking male contribution (Nogler 1984; Koltunow 1993). Thus, the seeds resulting from the apomictic process are clones that are fertile with identical genetic composition as the maternal parent except for possible mutations (Asker and Jeling 1992; Koltunow 1993; Koltunow and Grossniklaus 2003). However, the difference between apomixis and other mechanisms of vegetative reproduction is the formation of seeds within the female reproductive organ (Czapik 1994). Therefore, to understand the apomictic process it is necessary to know the sexual process better.

In different agamic complexes (with sexual and apomictic reproduction) harboring individuals with different ploidy levels, the genotypes that are diploid are usually sexual whilst the polyploids are generally apomictic. There is evolutionary evidence that hybridization and polyploidization precede apomixis, which might have helped to stabilize adapted genotypes through the transmission of unreduced female genotypes, thus facilitating the colonization of certain *habitats* (Appels et al. 1998; Carman 1997, 2001). This suggests that apomixis may have arisen via polyploidization or paleopolyploidization of ancestral parents that had a sexual mode of reproduction but contained divergent reproductive characteristics, during or after the Pleistocene. Studies with ancestral sexual relatives of *Tripsacum* and *Arennaria* indicated that the origin of apomixis is a result of heterozygosity and polygeny derived from floral asynchrony after genome doubling. In *Arabis holboellii*, studies using chloroplast haplotypes of individuals with different ploidy levels suggested that polyploidy arose independently and repeatedly (Sharbel and Mitchell-Olds 2001). Therefore, variation in the reproductive mode and structure of populations suggests that apomixis has a single evolutionary origin, with diverse expression of this trait.

Apomixis, usually associated with polyploidy, is genetically regulated and, in many species, is characterized as a dominant factor associated with one or more loci in a very complex Mendelian model. The understanding of the mechanisms that regulate apomixis is limited due to the polyploid nature of the plants, low fertility as a function of gene expression in sporophytic and gametophytic tissues, and factors such as epistatic interactions between genes, modifiers, segregation distortion and suppression of recombination, among others (reviewed by Ozias-Akins and van Dijk 2007). Several authors consider that the apomictic mechanism is not independent of the sexual one, and for that reason, the genes controlling apomixis would be the same involved in the sexual mode of reproduction, but with altered spatial and/or temporal regulation (Tucker et al. 2003; Koltunow and Grossniklaus 2003; Ozias-Akins and van Dijk 2007).

Advantages of Apomixis

According to Richards (1997), apomictic plants have the following advantages:

- Guarantee of the reproductive success even in the absence of pollination in environments of extreme climate conditions, except in the case of pseudogamous apospory and adventitious embryony that depend on the fertilization for seed development.
- Clonal reproduction equivalent to vegetative plant reproduction combined with the advantages of the seed, such as the absence of viruses, easy dispersion and dormancy.
- It avoids the cost of meiosis for the gamete formation (absence of recombination and segregation) and maternal energy would not be spent with ill formed zygotes since all the progeny is identical to the mother, who contributed with 100 % of its genotype, and not only with 50 % as in the case of allogamous sexual plants.
- Many apomictic plants can circumvent the cost of the male that does not produce pollen. However, this male sterility condition is not widely dispersed among the apomictic plants as the sterility genes are unable to disperse among the clones. Thus, many apomictic plants can act as male parents for fertilization of sexual plants, which can counteract the disadvantage of the energy used for pollen development.
- They can fix and spread genotypes extremely well adapted because the genotypes less adapted are eliminated by natural selection.

Disadvantages of Apomixis

The apomictic process has some disadvantages which, according to Richards (1997) are:

- Inability to prevent the accumulation of disadvantageous mutations to the reproductive success and adaptation by the absence of recombination and segregation.
- Inability to recombine advantageous features from mutations occurring in different individuals which would accelerate the evolution of the species towards environmental changes.
- Very narrow population niche.

Apomictic Mechanisms

The main events that characterize the apomictic mode of reproduction comprise the formation of the female gametophyte or embryo sac in the absence of meiotic reduction (apomeiosis); embryo development independent of fertilization (autonomous development or parthenogenesis), and endosperm development dependent (pseudogamy) or not on the fertilization. Cytologically, the process may be classified according to the origin and location of the cells (Fig. 2.2). Whenever its origin is in

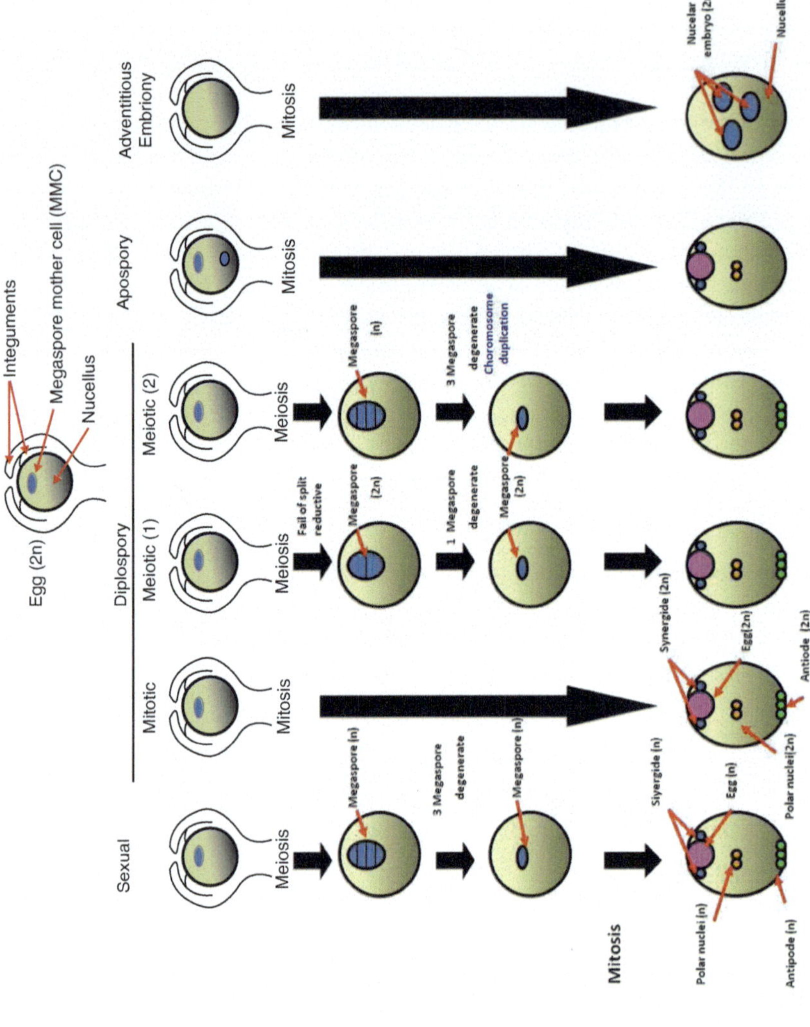

Fig. 2.2 Schematic drawing of apomictic and sexual process (Modified by MMG Karasawa based on Cavalli 2003)

the sporophyte tissue, the mechanism is designated adventitious embryony, whilst whenever the origin is the gametophyte the mechanisms are apospory and diplospory (Nogler 1984). The adventitious embryony can be characterized as a somatic embryogenesis, co-existing with the regular sexual process. In gametophytic apomixis, embryo develops autonomously (no male fertilization), originating from a meiotic unreduced embryo sac, with the sexual process affected in most cases.

The development of endosperm in apomictic plants can be autonomous (in the absence of fertilization) as in some species of *Compositae*, *Poaceae* and *Rosaceae* (Chaudhury et al. 2001) or pseudogamous, where the endosperm is the result of the fusion of the two polar nuclei (n + n) of the central cell with the male gamete cell, similar to what occurs in plants with the sexual mode of reproduction. Therefore, in apomictic pseudogamous plants, the endosperm is a bi-parental tissue, typically triploid as observed in *Brachiaria* sp. (Alves et al. 2001) and other apomictic plants.

1. **Diplospory**

The diplospory apomictic process (Fig. 2.3) can be identified by the differentiation of the megaspore mother cell into an embryo sac, which does not undergo a regular meiotic process (Nogler 1984; Koltunow 1993). It occurs through two mechanisms: mitotic or meiotic (Fig. 2.2).

- **Mitotic diplospory**

In the mitotic diplospory, the megaspore mother cell does not undergo meiosis and acts as a functional megaspore, which after going through three mitoses, gives rise to an unreduced embryo sac of *Antennaria* type (Fig. 2.3).

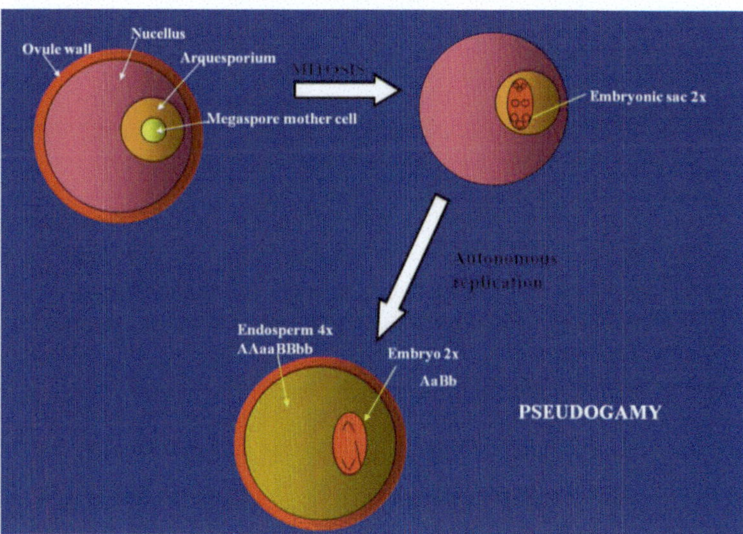

Fig. 2.3 Schematic drawing of the mitotic diplospory (Drawn by GCX Oliveira)

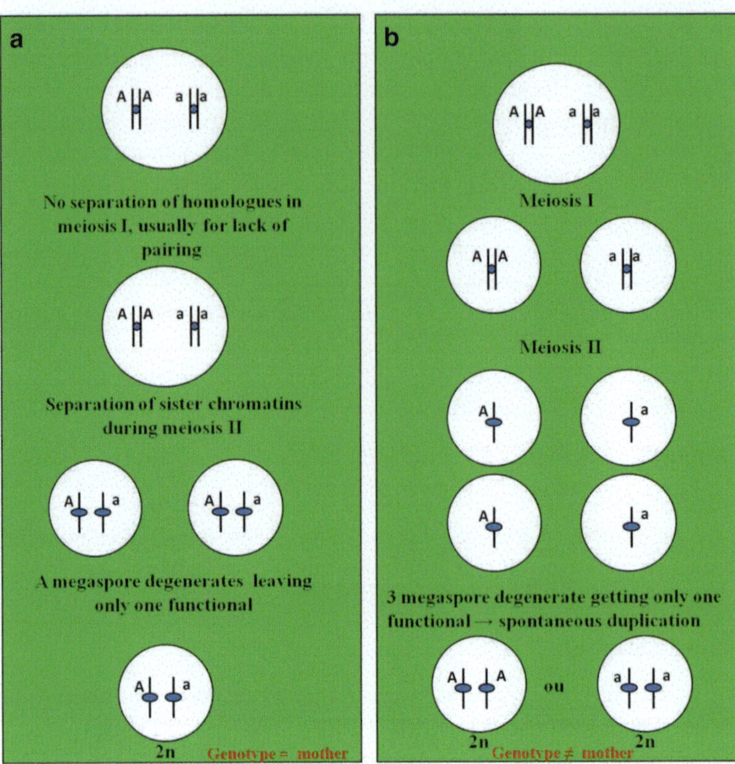

Fig. 2.4 Schematic drawing of the meiotic diplospory (Drawn by MMG Karasawa). (**a**) Fail of reductive division (**b**) spontaneous duplication of the chromosome number

- **Meiotic diplospory**

In the meiotic diplospory, the megaspore mother cell initiates meiosis and, due to a failure in the homologous chromosomes pairing and segregation during the prophase of meiosis I, it forms a restitution nucleus (Fig. 2.4). This cell proceeds normally to the second meiotic division, resulting in a dyad of unreduced cells. One of these degenerates, and the survivor, after undergoing three mitoses, forms the unreduced embryo sac of *Taraxacum* type. It is also possible that a restitution nucleus can be formed in the megaspore mother cell that undergoes the second meiotic division without cell division. After mitosis, this cell forms the unreduced embryo sac of *Ixeris* type.

The embryo develops from the egg cell and the endosperm from the nucleus of the central cell in the embryo sac. Diplosporous apomixis is found *in Allium*, *Ochna*, *Calamagrostis*, *Poa*, *Tripsacum*, *Taraxacum*, *and Ixeris*, among other genera (Asker and Jeling 1992; Koltunow 1993), and it is not normally associated with the sexual process.

2. **Apospory**

In apospory, an unreduced embryo sac of *Panicum* or *Hieracium* type is also formed. However, it differs from diplospory because the embryo sac originates from

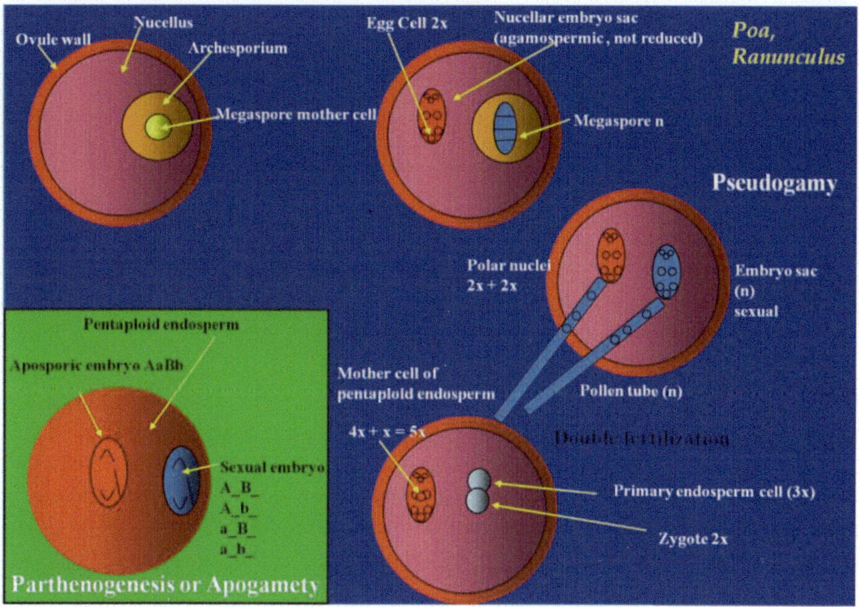

Fig. 2.5 Schematic drawing of the apospory (Drawn by GCX Oliveira)

cells of the nucellus, called aposporous initials and not from the megaspore mother cell. These cells contain a large nucleus and dense cytoplasm (Cavalli 2003) and were ultrastructurally characterized in *Brachiaria* spp. (Dusi and Willemse 1999; Araújo et al. 2000; Dusi 2001), as well as in other species. One or more aposporous initials undergo mitosis and give origin to the embryo sac, where the embryo will develop autonomously. *Hieracium* is another example of aposporous apomixis, with the embryo sac showing the cell arrangement and number similar to those observed in sexual plants (Koltunow 1993). Another example is *Brachiaria* that presents the *Panicum*-type embryo sac, with fewer cells and different organization from that found in the embryo sac formed in sexual plants (Araújo et al. 2000, 2005), which facilitates the morphological discrimination of reproductive mode (Fig. 2.5).

- **Facultative apomixis**

During the aposporous development, the megaspore mother cell can degenerate before or after the differentiation of the aposporous initials or it can undergo meiosis and form a reduced embryo sac. Thus, the sexual process can co-exist with the non-reduced embryo sacs originated from the aposporous initials within the same embryo sac. Consequently, some apomictic plants may also exhibit some sexual reproduction associated with apomixis, at different frequencies, and are called facultative apomictic plants. In these cases, a number of factors such as seasonal oscillation associated with the length of photoperiod during the inflorescence development and responses to day light length, intensity, temperature, type and level of soil fertility can cause a change in frequency of sexual embryos in apomictic plants (Koltunow

1993) as observed in *Dichanthium aristatum* (Knox 1967). In *Brachiaria* spp., the frequency of reduced embryo sacs associated with aposporous ones can vary from zero to 50 % (Lutts et al. 1994; Dusi and Willemse 1999; Valle et al. 1994; Valle and Savidan 1996, Araújo et al. 2004), depending on the species, accession and other undetermined factors.

3. **Adventitious embryony**

Adventitious embryony begins later and originates from mitosis of individual cells designated embryogenic initials, present in structural tissues of the ovary—nucellus or integument (Lakshmanan and Ambegaokar 1984). Several of these cells can differentiate into embryos, resulting in polyembryony (Asker and Jeling 1992). The most common type of adventitious embryony is the nucellar embryony (Fig. 2.6) and the embryos can co-exist within the same embryo sac with the zygotic embryo. However, the lack of endosperm in adventitious embryony generates a competition for reserves between sexual and apomictic embryos in development. The adventitious embryony is present mostly in plants of the family *Rutaceae*, *Liliaceae and Orchidaceae.* A classic example of nucellar embryony is *Citrus* (Koltunow 1993; Koltunow and Grossniklaus 2003).

4. **More than one type of apomixis**

Some plants have the simultaneous occurrence of different apomictic processes within the same ovule. In *Paspalum minus* the occurrence of aposporous and diplosporous embryo sacs within the same ovule has been described (Bonilla and

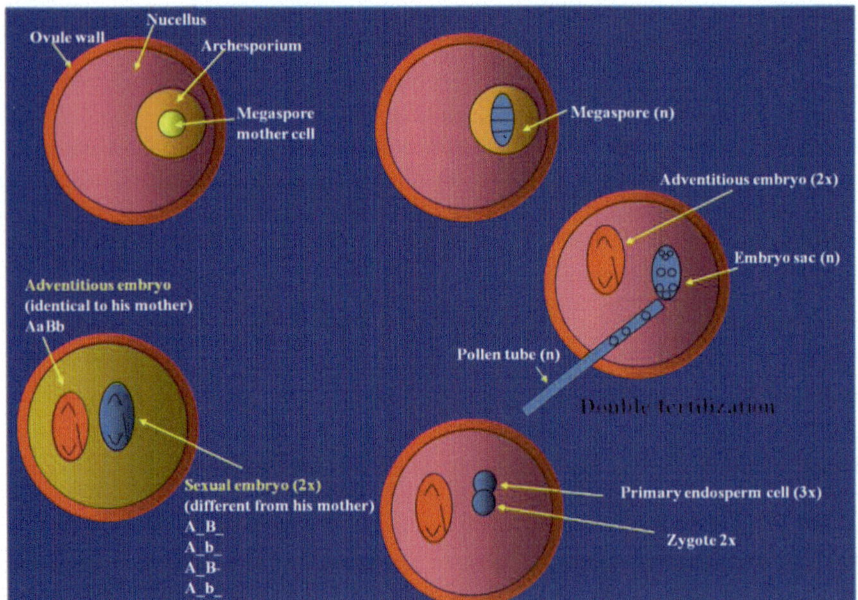

Fig. 2.6 Schematic drawing of adventitious embryony (Drawn by GCX Oliveira)

Quarin 1997). In the genus *Beta* and in Rosaceae, diplospory, apospory and adventitious embryony can be found together whilst apospory and adventitious embryony can co-exist in *Citrus* and *Hieracium* spp. (Koltunow and Grossniklaus 2003).

Pseudogamy

In some species, pollination is required to hormonally stimulate embryo development, even though the male gamete does not participate in the formation of the zygote. This process is called pseudogamy and has two main variants: sometimes one of the male nuclei has to fuse with the female polar nuclei generating the primary cell of the endosperm, and sometimes the endosperm develops autonomously, without any male contribution. This kind of apomixis is rare, being found in the family *Compositae*, occasionally in *Poaceae* and *Rosaceae*, and seldom in aposporous apomictic plants. Therefore, although apomictic plants do not require the male gamete for embryo development, in many cases the fertilization of the polar nuclei of the central cell in the embryo sac is necessary so that the endosperm can develop. This commonly occurs in *Brachiaria brizantha* (Alves et al. 2001). Moreover, there are plants in which the embryo development occurs prior to the fertilization of the polar nuclei—precocious embryogenesis.

Gene Control

Apomixis is not a process that receives only random stimulus of environmental and nutritional factors. Progeny analyses resulting from crosses between apomictic and sexual plants have shown that there is also a genetic control determining apomixis (Koltunow 1993). Analyses of apomixis inheritance are complicated due to the polyploid nature, existing compatibility among sexual plants and difficulty to define apomixis frequency in the progeny as a consequence of the segregation of the apomictic components (Koltunow and Grossniklaus 2003).

Initially, it was believed that the components of apomixis would be quantitative traits under polygenic control, with exception of apospory, for which there were already evidences of its control by a dominant gene in *Panicum* (Savidan 1989, 2000, 2001) and *Ranunculus* (Nogler 1984). Subsequently, it was considered that the control would be due to one or more non-recessive genes, and that apomeiosis and parthenogenesis would have independent control. Currently, there are evidences that genes controlling apomixis are usually dominant, observed in a simple form (single dose), and present in one or multiple loci in both monocots and dicots (Asker 1980; Asker and Jeling 1992). Indeed, apomixis segregates both in *Panicum* and in other members of the tribe Paniceae, as a single locus present in a chromosome region with low recombination frequency. However, analysis of the development of the embryo sac in *Panicum* indicated an earlier apomictic process relative to the sexual process (Savidan 2000), suggesting that the moment of the developing process activation is crucial and possibly results in the ectopic regulation of one or more genes. Thus, variations in the mechanisms could be merely consequences of different moments in the activation of apomixis. In *Panicum* spp. and *Ranunculus* spp. the locus that controls apospory co-segregates with the parthenogenesis locus, suggesting the existence of a single locus of simple or complex nature and dominant (Pupilli et al. 2001). Moreover, recent studies have shown that in *Panicum maximum* the components segregate separately (Kaushal et al. 2008).

Initially, it was also believed that apomixis in *Paspalum notatum* was recessive (Burton and Forbes 1960), but currently there are evidences that the inheritance is dominant with distortions in the segregation and recombination that are absent in the region that controls apospory (Pupilli et al. 2004; Martínez et al. 2003; Stein et al. 2003). In *Pennisetum*, segregation studies suggested that the region containing the locus of apomixis is dominant, heterozygous and disomically inherited (Ozias-Akins et al. 1998; Roche et al. 1999; Ozias-Akins and van Dijk 2007; Martínez et al. 2007). However, in both *Paspalum* and *Pennisetum*, the genomic region linked to apomixis does not show meiotic recombination. Also, in *Poa pratensis*, the mechanisms controlling apospory and parthenogenesis are dominant and heterozygous, but parthenogenesis is contingent on apospory (Albertini et al. 2001a, b, 2004; Matzk et al. 2005). The great variation in the expression of parthenogenesis suggests that it is under the control of a complex loci or under the effect of modifiers yet undetermined. The model currently accepted is four genes, including the apospory initiator and inhibitor and the parthenogenesis initiator and inhibitor. Moreover, the presence of a fifth element is considered to be regulating the megaspore development (Albertini et al. 2001a; Matzk et al. 2001, 2005; Porceddu et al. 2002).

In *Taraxacum*, there are three independent loci with dominant action with simple genotype; two are associated to diplospory and parthenogenesis and the third has not yet being determined. Furthermore, independent genetic control was observed during the formation of autonomous endosperm, as well the occurrence of a barrier against fertilization, suggesting the involvement of a fourth apomictic element (Záveský et al. 2007).

In *Brachiaria*, crosses between tetraploid *B. ruziziensis* (with sexual reproduction) and *B. brizantha* (apomictic) suggested that the inheritance of apomixis is simple with a dominant allele (Miles and Escandon 1997). Also, studies of gene expression by Leblanc et al. (1995a, b) and Albertini et al. (2004) did not show genomic regions that segregated on available genetic maps; therefore, these studies could not show the linkage group involving genes controlling the reproduction. Recently, Rodrigues et al. (2003) cloned and sequenced transcripts during the development of an apomictic and sexual ovule of *B. brizantha*, where 11 clones showed differential expression in the development stage or the genotype.

In *Hieracium*, an aposporous plant, different loci are associated with the initialization process of apomixis, such as the number of embryo sacs formed and the development pathway, suggesting the occurrence of an epigenetic regulation (Koltunow et al. 1998, 2000; Bicknell et al. 2000; Bicknell and Koltunow 2004). Catanach et al. (2006) found that in *H. caespitosum*, two major loci control apomixis; one controls the events associated with apomeiosis and the other, those associated with the formation of a barrier preventing fertilization.

Recent research (Ozias-Akins and van Dijk 2007; Matzk 2007; Noyes et al. 2007) reported that in *Erigeron*, the inheritance of apomixis involves regular

Mendelian segregation and that parthenogenesis is dependent on diplospory. Studies in *Hypericum* show the dominance of the character, and in *Parthenium*, the inheritance control is independent of the components (Barcaccia et al. 2006).

Current data indicate that apomixis is usually dominant and segregates as if controlled by one to three genes, with some exceptions. Moreover, the genetic control is complex and is under epigenetic regulation.

Gene Isolation

Strategies for the isolation of genes involved in apomixis currently consist of (1) introgression of the character from apomictic varieties close to those of interest; (2) comparison of genes differentially expressed during development in populations with natural sexual and apomictic mode of reproduction; (3) analysis of induced mutants that have lost or increased apomictic character; and (4) mutagenesis in sexual species to obtain apomictic characters. Experiments aiming to introgress apomixis in maize (Savidan 2001) and *Pennisetum*, using apomictic wild relatives, were unsuccessful. There was also an effort to induce apomixis in rice by mutagenesis (Khush 1994), without success. In *Arabidopsis thaliana* several genes related to embryogenesis have been isolated, including genes capable of producing endosperm or initiate the formation of embryo independent of fertilization, a component of apomixis. Some of these genes belong to the *fis* group (*fertilization-independent seed*), including *MEDEA* (*MEA*), *FIS2* and *fertilization-independent endosperm* (*FIE*), and their regulation is through genomic *imprinting* (Grossniklaus et al. 1998, 2001; Luo et al. 1999, 2000; Kinoshita et al. 1999; Vielle-Calzada et al. 1999; Koltunow and Grossniklaus 2003; Rodrigues and Koltunow 2005; Spillane et al. 2000; Yadegari et al. 2000).

Several other mutants containing genes involved in the development of functional male and female spores, embryo sac, egg cell, parthenogenesis, embryogenesis and endosperm have been identified (reviewed by Koltunow and Grossniklaus 2003); among others, *feronia* using the detection strategy of *enhancer* (Huck et al. 2003), mutants with abnormal pattern of development of apomixis using γ ray and insertional mutagenesis in *Hieracium* spp. (Bicknell et al. 2001), *Arabidopsis* mutants using mutagenesis with transposon to generate embryo sacs containing multinucleated cells probably originated from functional megaspores. The knowledge of the mechanisms of apomictic reproduction has been expanding, with studies on the inheritance, genetic mapping, isolation and induction of mutants, together with the characterization of genes involved in the reproductive process (*SG-1, APOSTART, SERK, AINTEGUMENTA, BABYBOOM, KNUCKLES, SPOROCYTELESS (SPL/NOZZLE)* and meiotic mutants) and will, hopefully, contribute to enable the introduction of the apomictic character through genetic engineering in crops with sexual reproduction, a tool of unlimited interest for agriculture.

Recently, Curtis and Grossniklaus (2008) obtained mutants of the *fis* group with two distinct phenotypes: if fertilized, the seeds generated showed an aberrant proliferation of the embryo and endosperm, and eventually, aborted the zygote, and in the absence of fertilization all mutants began to form the autonomous endosperm from the polar nuclei and developed the embryo. This study demonstrated that in *Arabidopsis*: crossing wild plants WT promoted the development of normal embryo and triploid endosperm; the mutant type *MEA/mea* developed embryo and diploid endosperm autonomously; mutants of the type *msi1* developed, autonomously, non-viable haploid embryo through irregular parthenogenesis and diploid endosperm; crosses using WT as maternal parent and pollen from the mutant *CDK;1/cdk;1* generated the development of a mutant seed and remnant diploid endosperm, showing the abortion of the embryo at an early globular stage; crosses using *GLC/glc* as the mother with WT as the father generated the development of mutant seeds without endosperm in the absence of fertilization of polar nuclei; crosses using maternal parent *mea/mea* × *CDK;1/cdk;1* generated a small but viable mutant seed, containing diploid endosperm and embryos.

2.3 Sexual Reproduction

Sexual reproduction is the process whereby the union of male and female gametes occurs in order to produce the zygote. It is based on two principles: to generate variability by means of *recombination*, *segregation* and *sexual fusion* (syngamy), and to promote *genic migration* by means of the exchange and incorporation of genes (Richards 1997).

2.3.1 Why Sex?

This is one of the most controversial matters in Biology. The cost of sexual reproduction in relation to asexual reproduction is twofold, because it involves the search for mates, the destruction of coadapted genic complexes and costs with the male function (Maynard Smith 1971). Sexual reproduction requires two individuals (a male and a female) to produce a number x of progeny, being x the average number that a mother can generate. The male contributes only the gametes to the process. On the other hand, in asexual reproduction two individuals produce $2x$ progeny (Fig. 2.7, where $x=2$).

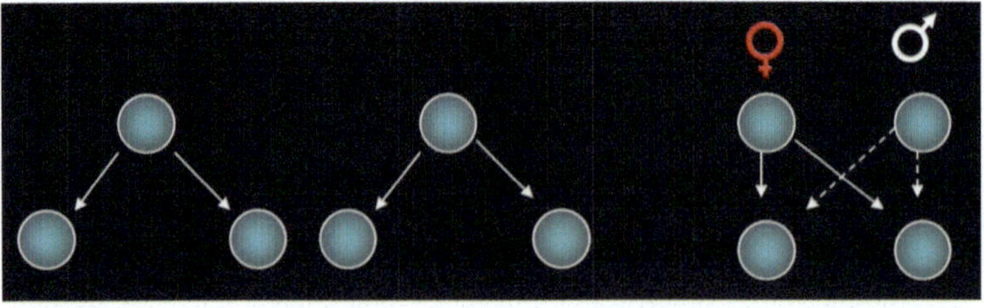

Fig. 2.7 Maternal and paternal participation in progenie formation (Drawn by GCX Oliveira)

2.3.2 Advantages of Sex

Understanding the advantages of sex requires analyses based on the individual's reproductive success, and not on "the good of the species"; thus, the "generation of fuel for evolution" argument is inadequate. As sex arose in unicellular beings, the scenarios for its evolution must be situated in a unicellular world (Fig. 2.8).

Many unicellular species evolved a sporulation mechanism as an adaptation for resistance to occasional stressing environmental conditions. The sporulation genes undergo mutations for several generations without any selection pressure in the absence of stress. When the stress eventually occurs, a spore is produced (as a resistant form), which exposes the genes to selection. Mutations allowing cell fusion gave rise to a new life strategy—diploidy—in which deleterious mutations in the sporulation genes are compensated for by wild alleles present in the homologous chromosome (Fig. 2.9).

Thus, meiosis would have evolved because of the advantage it confers to the cell in preventing the ploidy level to increase indefinitely. However, the fusion of genetically identical cells does not bring good results, for the mutations are identical in both homologous chromosomes and are not compensated for. Conversely, fusions between genetically different cells are likely to produce spores with at least one perfect copy of each gene, creating variability as a byproduct. Plants, which inherited meiosis and cell fusion from unicellular organisms, display mechanisms that prevent autozygosis, i.e., the presence of genes identical by descent in the

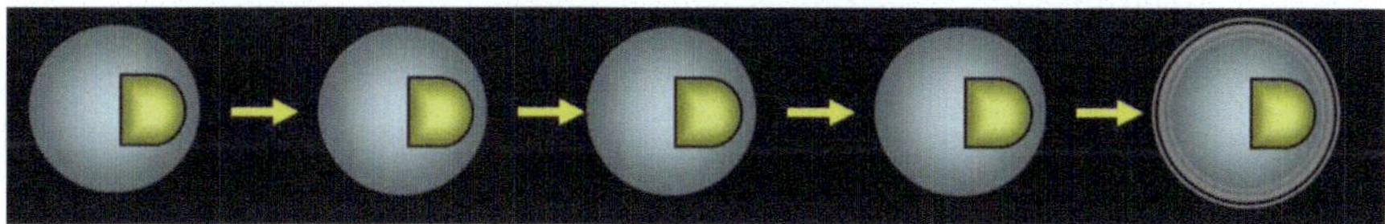

Fig. 2.8 Eschematic drawing of sexual reproduction (Drawn by GCX Oliveira)

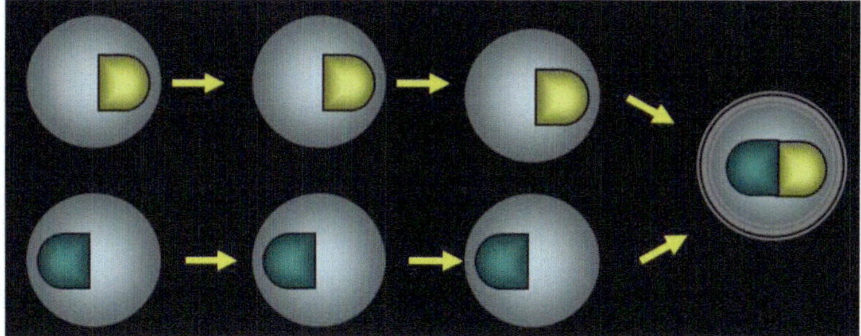

Fig. 2.9 Schematic union of haploid cells to form a diploid, precursor of fertilization (Drawn by GCX Oliveira)

same diploid cell. In plants, the main phenomenon associated to the formation of autozygotes is autogamy, and many mechanisms evolved in this group to force the occurrence of crossings or increase its probability.

2.3.3 Disadvantages of Sex

Sexuality shows the following disadvantages (Richards 1997):

- Sexual mothers spend resources with genetically variable offspring which may be better adapted to new environments, but may as well be little adapted to the niche where they live. The offspring of an asexual mother, on the other hand, will be all identical to her and thus as well adapted to their environment as she is.
- Within a sexual population, a mutant individual that lost the female fertility but kept the male fertility will contribute the mutant allele through male gametes, but will not be able to receive normal female fertile alleles through male gametes from the population. This determines an Evolutionary Stable Strategy (ESS) which leads to the spreading of the phenotype across the population.
- Obligatory sexuality may be disadvantageous in the absence of pollinators, because the need of crossing may reduce reproductive efficiency.
- Marginal environments, which present barely tolerable conditions to the species, are often more homogeneous than the rest of the species ecological range. In such environments, the genetic invariability guaranteed by asexuality can be more advantageous. Moreover, extreme conditions are more likely to damage the sexual organs than those used for asexual reproduction.

2.3.4 Gametogenesis and Fecundation

Angiosperms Life Cycle

The life cycle of the Angiosperms is composed by the alternation of two types of generation, the gametophytic generation (haploid phase) and the sporophytic generation (diploid phase) (Fig. 2.10). Gametophyte formation involves few cells located in the flowers, which are the sexual organs. The establishment of the gametophytic phase requires the specification and differentiation of the gametes in the anther and in the ovule. The gametophytic generation begins with the differentiation of the megaspore and microspore mother cells, while the sporophytic generation begins with fertilization (Maunseth 1995), which follows the transfer of the male gametophyte from the anther to the stigma present in the gynoecium. Thereafter, the male gametophyte develops into the pollen tube, which takes both male gametes to the embryo sac, where the double fertilization will take place. One of the gametes fertilizes the egg cell, while the other unites to the polar nuclei. After the fertilization process, a number of embryogenic events give rise to the embryo, which pins

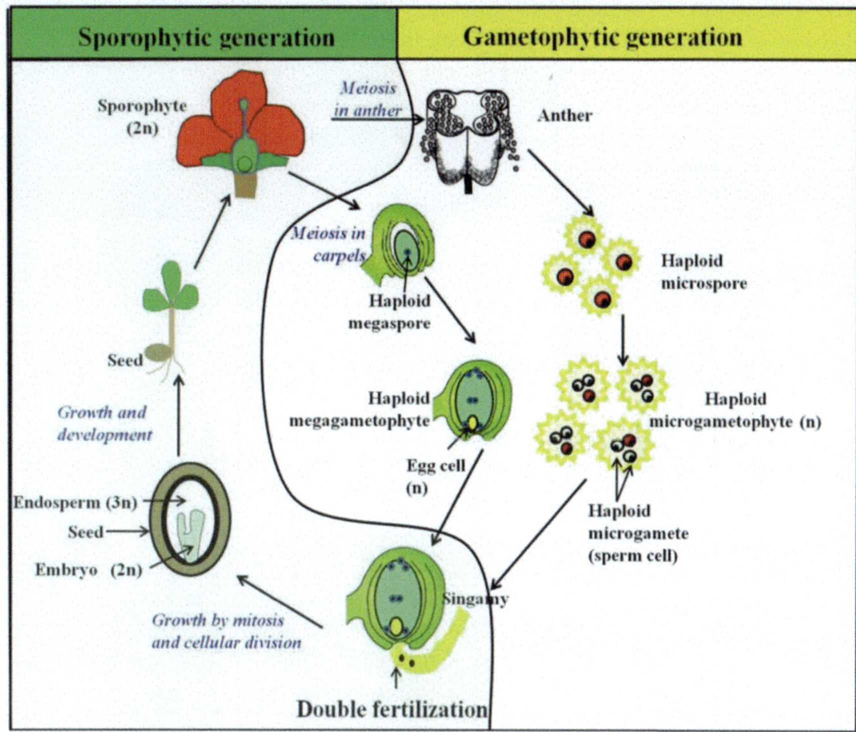

Fig. 2.10 Angiosperms life cycle (Drawn by MMG Karasawa)

down the beginning of the sporophytic phase (Drews and Yadegary 2002). However, little is known about the genetic basis and the molecular mechanisms that regulate gametogenesis in angiosperms (Estrada-Luna et al. 2002).

Gametophytic Generation

- **Formation of embryo sac**

The female gametophyte, also known as megagametophyte, develops within the ovary in the interior of the flower (Drews and Yadegary 2002). The female gametophytic generation starts out when the megaspore mother cell enters meiosis (Fig. 2.11). The process comprises meiosis I and II, which result in the production of four haploid cells. These four cells are arranged along the chalazal-micropylar axis. Thereafter, three out of the four megaspores degenerate, and only that located near the chalaza survives (Zanettini and Lauxen 2003). The second stage of the process, in its more typical variant, comprises a sequence of three cariocineses (mitoses), which result in eight nuclei in the embryo sac (three antipodes, two synergids, two polar nuclei and one egg cell). Once they are formed, the antipodals

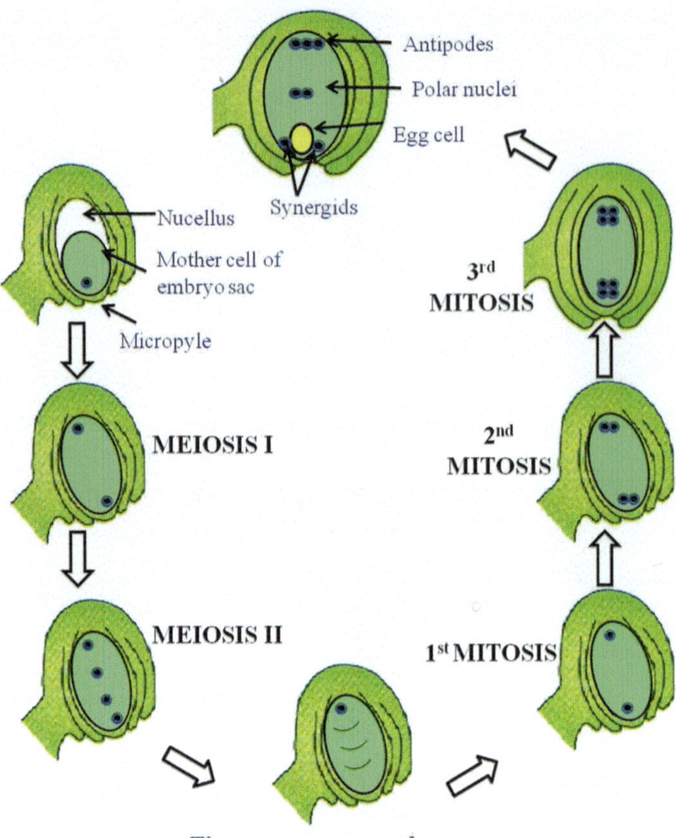

Fig. 2.11 Scheme of embryo sac formation (Drawn by MMG Karasawa)

migrate to the apical region, near the chalaza, through which all the necessary nutrients are transferred. The polar nuclei reach the median region of the embryo sac by the time the synergids and the egg cell are lodged near the micropyle, wherethrough the pollen tube penetrates carrying the reproductive nuclei, which will affect the double fertilization. The analysis of the female gametophyte is important because it is an integral part of the plant life cycle and is essential for the formation of the seed. Moreover, it signals for and aids the pollen tube during the double fertilization, and expresses the genes that control embryo and endosperm development (Drews and Yadegary 2002).

- **Genetic control of embryo sac formation**

Studies performed by Drews and Yadegary (2002) in *Arabidopsis* identified the genes *AGAMOUS*, *APETALA* and *BELL1*, responsible for the development of the ovary. *APETALA* and *BELL1* also control the formation of the chalaza and the integument. The genes *HUELLENOS*, which affects both ovule and funiculus growth, and *AINTEGUMENTA* presented positive regulation of the marginal tissue, placenta and ovule formation. The identified mutants which affect the formation of embryo sac were distributed into five categories: (1) those that affect the initial stages and cause fail on the progression from the uninucleate stage of the megaspore mother cell; (2) those that provoke a defect in the nuclear divisions from the binucleate to the octonucleate stages, with consequences for the number, position and arrangement of the nuclei in the development stages 2–5, and cause cellularization to fail; (3) those that affect the cellularization process; (4) those that affect the fusion of the polar nuclei; (5) those whose phenotype is typically wild and apparently have no effect on gametogenesis. Punwani and Drews (2008) reported that the absence of the synergid filiform apparatus in FERONIA prevents the pollen tube from bursting thus rendering double fertilization impossible.

- **Pollen grain formation**

The development of the male structure requires the formation of the stamen and the differentiation of the tissues that build the anther (Ma 2005). The male gametophyte, or pollen grain, develops within the anthers (Drews and Yadegary 2002), which are composed of four pollen sacs, fused and linked to the filament (Fig. 2.12).

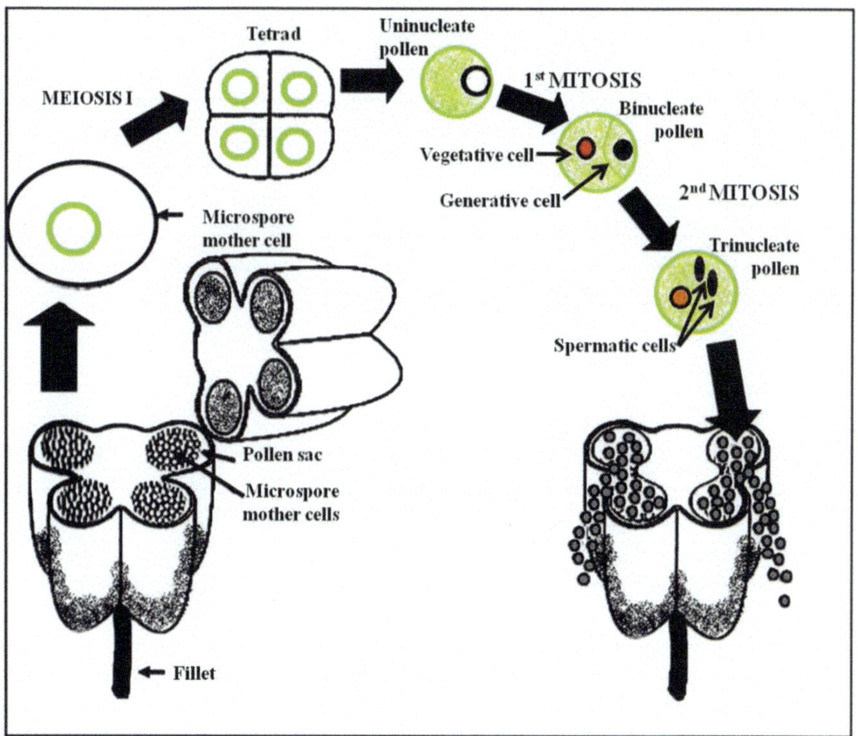

Fig. 2.12 Scheme of pollen grain formation (MMG Karasawa based on Zanettini and Lauxen 2003)

Fig. 2.13 Scheme of pollen wall (Karasawa et al. 2006)

Within each sac there are the microspore mother cells, surrounded by a tissue called tapetum, which nourishes the microspores during development and maturation. The microspore mother cells are diploid cells that undergo meiosis and produce four haploid cells (microspores), which initially remain united in a tetrad encapsulated by a callose wall. As the tetrad develops, the enzyme callase is secreted and digests the callose wall, releasing the microspores (Zanettini and Lauxen 2003). The microspores undergo two consecutive mitotic divisions and then differentiate (Goldberg et al. 1993) and become the mature pollen grain which is later released as the anther dehiscence (Ma 2005).

The pollen wall is composed of an internal layer known as intine, surrounded by an external layer called exine. Intine is similar in composition to most primary cell walls, and is composed basically of cellulose (Fig. 2.13). Exine, on the other hand, is derived mainly of material deposited by the tapetum and of other substances such as flavonoids and lipids. Knowledge about the different layers that compose the pollen grain wall is very important because they provide protection against desiccation and retain in the columella chambers the factors (chemicals) that control the self-incompatibility systems (SI). These chemicals are released in the stigma during pollen grain germination; in case of compatibility, the pollen tube grows and fertilization of the egg cell occurs, leading to the zygote; otherwise, the pollen tube dies and no zygote is formed.

Sporophytic Generation

The compatible pollen grain deposited on the stigma surface germinates and emits a pollen tube that is directed by two chemotaxis systems. One directs the pollen tube towards the micropyle and the other directs the penetration in the micropyle (Márton and Dresselhaus 2008). The former is influenced by the synergids, while the latter is believed to be controlled by two cells of the female gametophyte yet to be identified. The union of the male gamete to the female gamete, which results in the formation of the zygote, gives rise to the sporophytic generation (Fig. 2.14).

The newly formed zygote undergoes embryogenesis and will constitute the seed together with additional maternal tissues. The basic body plan of the sporophytic stage is established during the embryogenesis in a similar way in all angiosperms (Fig. 2.15), with differences only in the precision of the cell division patterns, extension of endosperm development and extension of the development of root and shoot

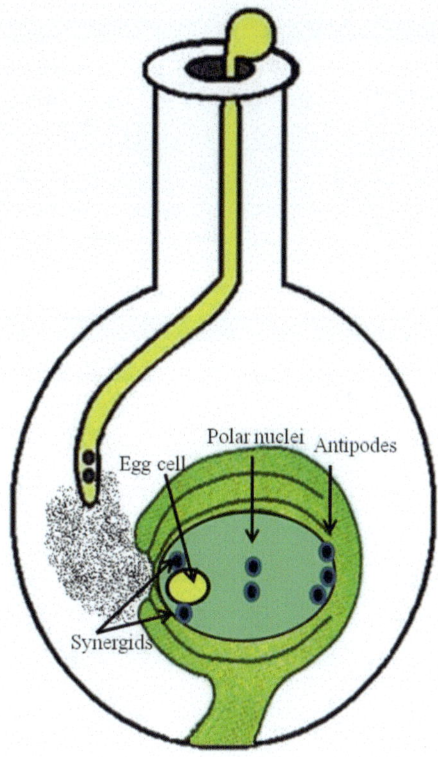

Fig. 2.14 Germination of compatible pollen grain on stigma (Drawn by MMG Karasawa)

meristems (Dornelas 2003). The angiosperm zygote normally divides transversely forming both an apical cell, which will give origin to the embryo, and a basal vacuolated cell, which will give origin to the structure known as the suspensor (Yeung and Meinke 1993).

Once the apical-basal polarity is established, a series of longitudinal and transversal divisions take place in the apical cell until the embryo in the globular stage is formed. The emergent shape of the embryo depends on the regulation of the division and expansion planes. The division planes in the outermost layer, called protodermis, are restricted to this layer, making it unique. The radial pattern emerges in the globular stage with the initiation of differentiation of the three tissue systems: lining, filling and vascular. The lining tissue (epidermis), derived from the protodermis, provides the external protecting layers. The filling tissue makes up most of the tissues under the protodermis (i.e., the cortex and the parenchyma). The vascular tissue originates the procambium (xylem and phloem), which works both as a mechanical frame for the plant and in sap transport. In dicots, the globular shape of the embryo is lost with the development of the cotyledons and is replaced by the heart-shape stage, which is followed by the torpedo stage. The transition from radial symmetry (typical of the globular phase) to bilateral symmetry (typical of the heart phase) seems to be mediated by hormones, especially auxin.

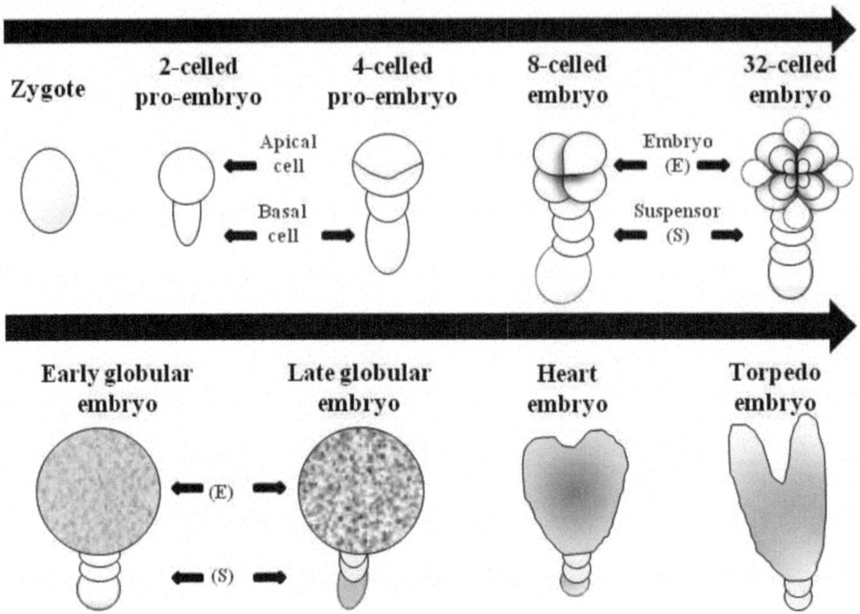

Fig. 2.15 Development stage of sporophytic embryogenic phase (Drawn by MMG Karasawa)

Both shoot and root apical meristems are formed by groups of cells that retain the multiplicative ability and persist during the post-embryonic stage, giving rise to most of the sporophytic body. The shoot meristem, in some species, is partially derived from the hypophysis, while all the other parts of the body come from the embryo. Studies on developmental genetics indicate that both root and shoot formation is controlled in an independent way (Dornelas 2003). Different mutants that affect each of the phases have been identified in maize and *Arabidopsis*, indicating an effective gene control. Among them, the mutant *GNOM/EMB30* was found to affect the apical-basal polarity of the embryo. *GNOM* prevents the zygote from elongating in the same way as the wild type does, because the first division seems to be symmetrical. The *GNOM* zygote cannot form a root either, and has a reduced apical structure. Moreover, the mutation in the *GURKE* gene suppresses the apical part and leads to the formation of a green mass in the embryo's stead. Mutations in the *FACKEL* gene (*FK*) reduce the hypocotyl, which produces seeds with the cotyledon attached to the root (Chaudhury et al. 2001). Independent control has also been shown in the maize mutant *dek23* and in the *Arabidopsis* mutant *stm* (shoot-meristemless). Both mutants determine the presence of root meristem and the absence of shoot meristem. On the other hand, mutations in the gene *HOBBIT* of *Arabidopsis* affect the development of the hypophysis, preventing the formation of the root meristem (Dornelas 2003). This is a small sample of the mutants related to the embryogenesis control.

Life Cycle of Gymnosperms

The life cycle of the Gymnosperms is composed of the gametophytic and sporophytic phases (Fig. 2.16). This was the first plant group to present strobili, which are incomplete flowers, and do not form ovaries. For this reason, they produce naked seeds, without fruits. The strobili, structures involved in the reproduction of the Gymnosperms, are located in the modified terminal branches and are composed of fertile leaves called sporophylls, because they produce spores. There are two types of sporophylls: the microsporophyll, which produces microspores, and the

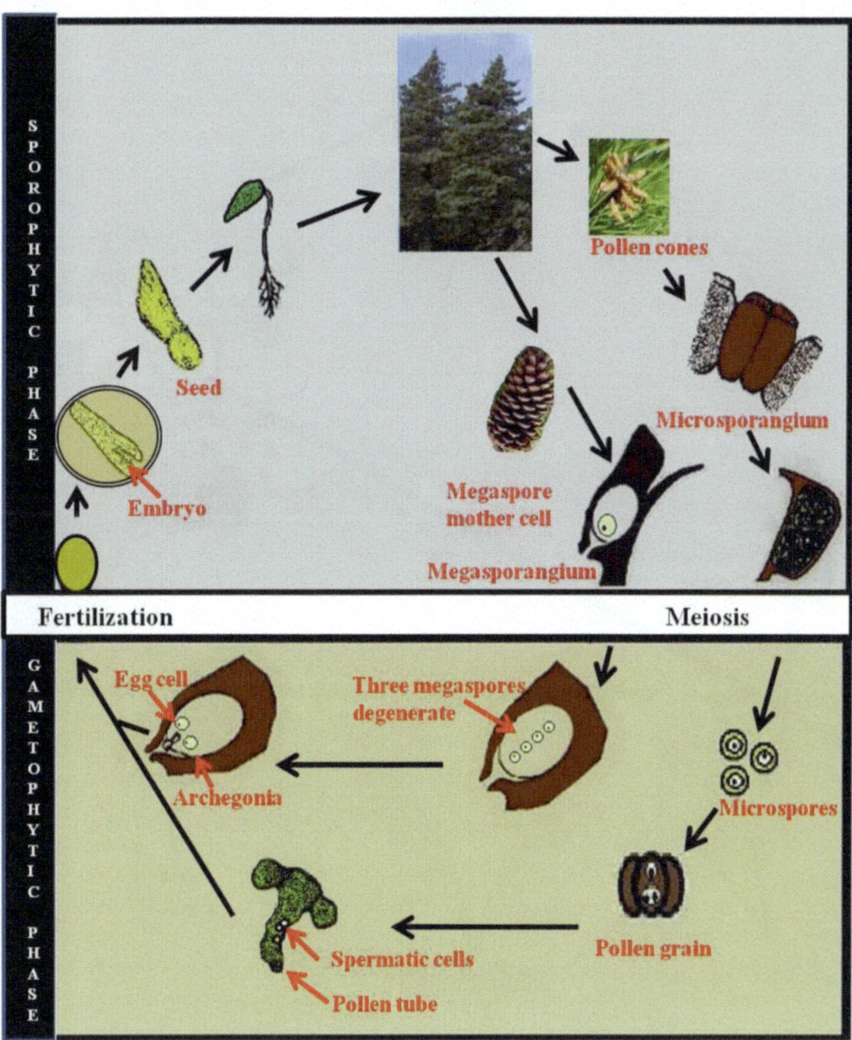

Fig. 2.16 Gymnosperms life cycle (Drawn by MMG Karasawa)

megasporophyll, which produces megaspores. Two microsporangia develop on each microsporophyll, and many microspores are produced within a microsporangium. The microspores initiate the formation of the pollen grain through mitoses and differentiation still inside the microsporangium.

Life Cycle of Pteridophytes

The haploid life phase (gametophytic) is small and short-lived, while the diploid phase (sporophytic) is larger and long-lived (Fig. 2.17). The cycle begins with the formation of haploid spores which fall on the ground and germinate. Then the spore

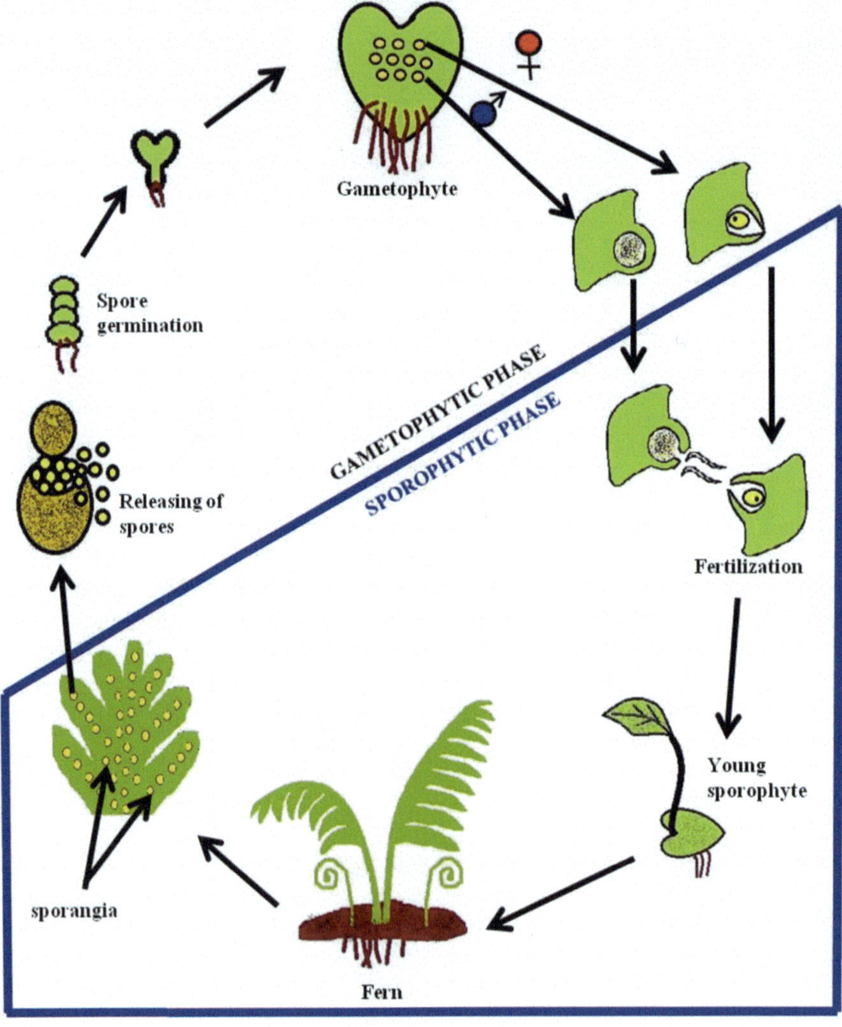

Fig. 2.17 Life cycle of pteridophytes (Drawn by MMG Karasawa)

gives rise to a small heart-shaped blade, the prothallus, where the reproductive organs, the antheridia and the archegonia, are produced. Here the male gamete, produced in the antheridium, needs a drop of water to reach the female gamete located in the archegonium. Upon the fusion of the gametes, starts the diploid life stage, the sporophyte, formed by roots, shoot and leaves. On the lower face of some leaves, groups of sporangia, the sori, are formed, and there the diploid "spore mother-cells" undergo meiosis, producing haploid spores. These spores germinate and produce a new prothallus, restarting the cycle.

Life Cycle of Bryophytes

In lower plants the gametophytic phase is dominant (Drews and Yadegary 2002) and starts with meiosis, which produces the haploid spores (Fig. 2.18). These germinate and grow into a structure called protonema (n), which develops organs

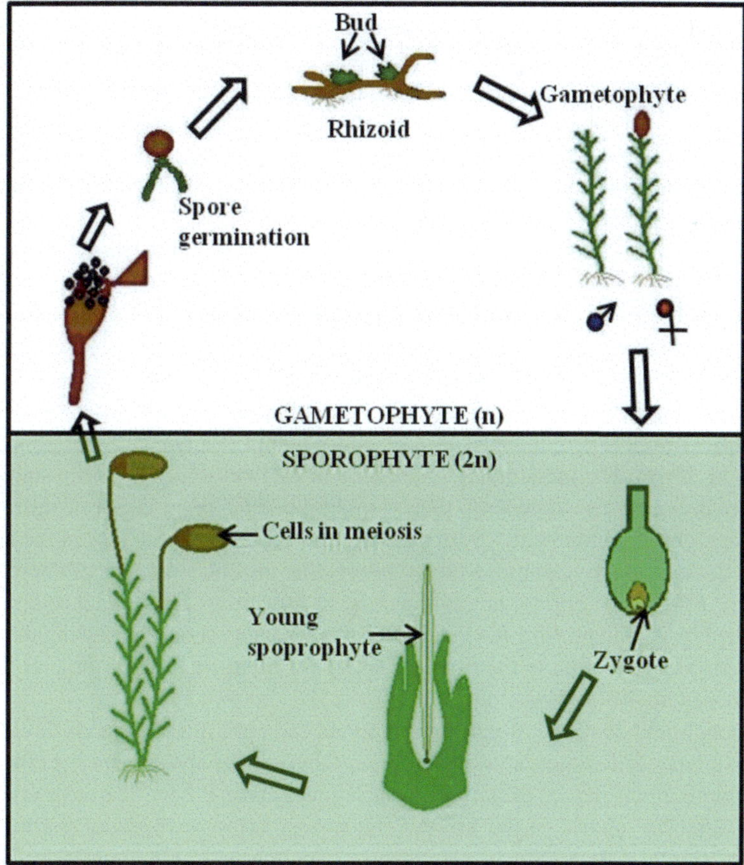

Fig. 2.18 Life cycle of bryophytes (mosses) (Drawn by MMG Karasawa)

resembling leaves, the phylloides (n). At their ends, these phylloides form the male (antheridium) and the female (archegonium) reproductive organs. In bryophytes, the phenomenon known as dioecy (separate sexes in male and female plants) occurs. The antheridium produces a flagellate gamete that depends on water to swim to the female gamete, which is found in the archegonium. After the union of the gametes (fecundation), a short-lived, diploid structure—the sporophyte or sporophytic generation—appears and produces a capsule (2n) at its end, where the spores are produced by meiosis and the cycle restarts. The sporophyte develops on the female gametophyte and is never found in isolation. In lower plants (bryophytes) the gametophytic phase is the conspicuous, dominant generation, while the sporophytic phase is nutritionally dependent and less complex.

2.3.5 Genetic Control of the Reproductive Organs

Lately, genetics and molecular biology studies have demonstrated that the mechanisms controlling reproductive development are much conserved in plants, even between angiosperms and gymnosperms (for a review, see Dornelas and Dornelas 2005). The first step in floral development is the transition from the vegetative to the reproductive phase, during which the vegetative meristem, which produces only leaves, starts to produce also floral meristems. These are the specialized structures which will develop into flowers in angiosperms. In gymnosperms, there is no production of flowers, but rather of male and female cones (strobili).

During the reproductive phase of angiosperms, the genes for identity of the floral meristem, most of them coding for transcription factors, promote the initiation of individual flowers. In the model plant *Arabidopsis thaliana* (an angiosperm of the family Brassicaceae), the major genes for identity of the floral meristem are *LEAFY* (*LFY*) and *APETALA1* (*AP1*) (Mandel et al. 1992; Weigel 1998). These genes are not only necessary for flower initiation, but are sufficient for flower induction when their superexpression is induced in transgenic plants (Weigel and Nilsson 1995; Peña et al. 2001). The induction of expression of *LFY* orthologs has been studied in detail in angiosperms. In general, the expression of *LFY* is initially low during the vegetative phase, but increases with age. Expression levels are higher in the beginning of the reproductive phase, which suggests that the concentration of the product of gene *LFY* may be critical in the transition to flowering. This evidence has been confirmed by the demonstration that in transgenic plants super expressing the gene *LFY* leads to a reduction in the time necessary for the formation of the first flower (Blázquez et al. 1997; Peña et al. 2001).

Although a wide range of genetic mutations can alter the process of flower formation, relatively few genes have been found that are involved in the specification of the floral organs *per se*. Mutations in such genes cause homeotic transformations in two adjacent whorls of the flower. The two outermost whorls of the mutants *apetala2* (*ap2*) of *Arabidopsis*, for instance, have carpels and stamens instead of sepals and petals, respectively. Mutations in the genes *APETALA3* (*AP3*) or *PISTILLATA* (*PI*) of *Arabidopsis* provoke the substitution of sepals for petals and of carpels for stamens. Finally, in the mutant *agamous* (*ag*) of *Arabidopsis*, the two innermost whorls are altered: the stamens are transformed into petals and the carpels into sepals (Coen and Meyerowitz 1991; Meyerowitz et al. 1991; Ma 1998). The modifications in the characteristics of the floral organs of the mutants just described suggest a simple combinatorial model for the determination of the identity of these organs (Coen and Meyerowitz 1991). According to this model, called "ABC Model", the genes responsible for identity are active in three overlapping regions, each one comprising two adjacent whorls.

Due to the superposition of the expression regions of each gene, a single gene combination specifies the identity of each whorl. If activity region B (which requires the expression of genes *AP3* and *PI* in *Arabidopsis*) is absent, both whorls 1 and 2 will be specified only by activity region A (*AP2* in *Arabidopsis*) and will contain sepals (Fig. 2.19).

Likewise, in this case, whorls 3 and 4 will be specified by activity region C (*AG* in *Arabidopsis*) and will contain carpels. In order to explain the phenotypes of mutants *ap2* and *ag*, it is necessary to add to the model the prediction that activities A and C are mutually antagonistic. That is, in a mutant for type-A genes, the action of C is extended to the four whorls and, analogously, in a type-C mutant, the activity of A is expressed in the four whorls.

The ABC Model, created for explaining the phenotypes of simple mutants, passes a final test: it predicts precisely the phenotypes of double mutants. For instance, the model predicts that, should B and A functions be removed, C would define the identity of the four whorls, which would develop into carpels. Indeed, all the whorls of the double mutant *ap2ap3* contain only carpels (Meyerowitz et al. 1991). Similarly, if activities B and C were absent, A should define the identity of all floral organs. As predicted by the model, sepals develop in all whorls of double mutants *ag pi*. The phenotype of this double mutant shows several additional concentric whorls (all composed of sepals), due to the effect of mutation *ag* of suppressing the determination of the floral meristem.

What would happen if both function A and C are removed?

Well, the activity B alone would define the identity of whorls 2 (petals) and 3 (stamens), but none of the activities identified would be present in whorls 1(sepals) and 4(carpels) (Fig. 1.15). The model does not make any obvious prediction about the resulting phenotype, because none of these states occur in any of the wild type flower whorls. Function B is associated to the formation of petals and stamens; thus, B is expected to cause the production of organs intermediate between petals and stamens in the absence of A and C. In fact, whorls 2 and 3 in the flowers of the double mutant *ap2ag* are occupied by staminoid petals. Whorls 1 and 4 of this double mutant contain leaves. Likewise, in the triple mutant *ap2ap3ag*, where functions A, B and C are deactivated, the flowers are formed by leaves organized in several concentric whorls.

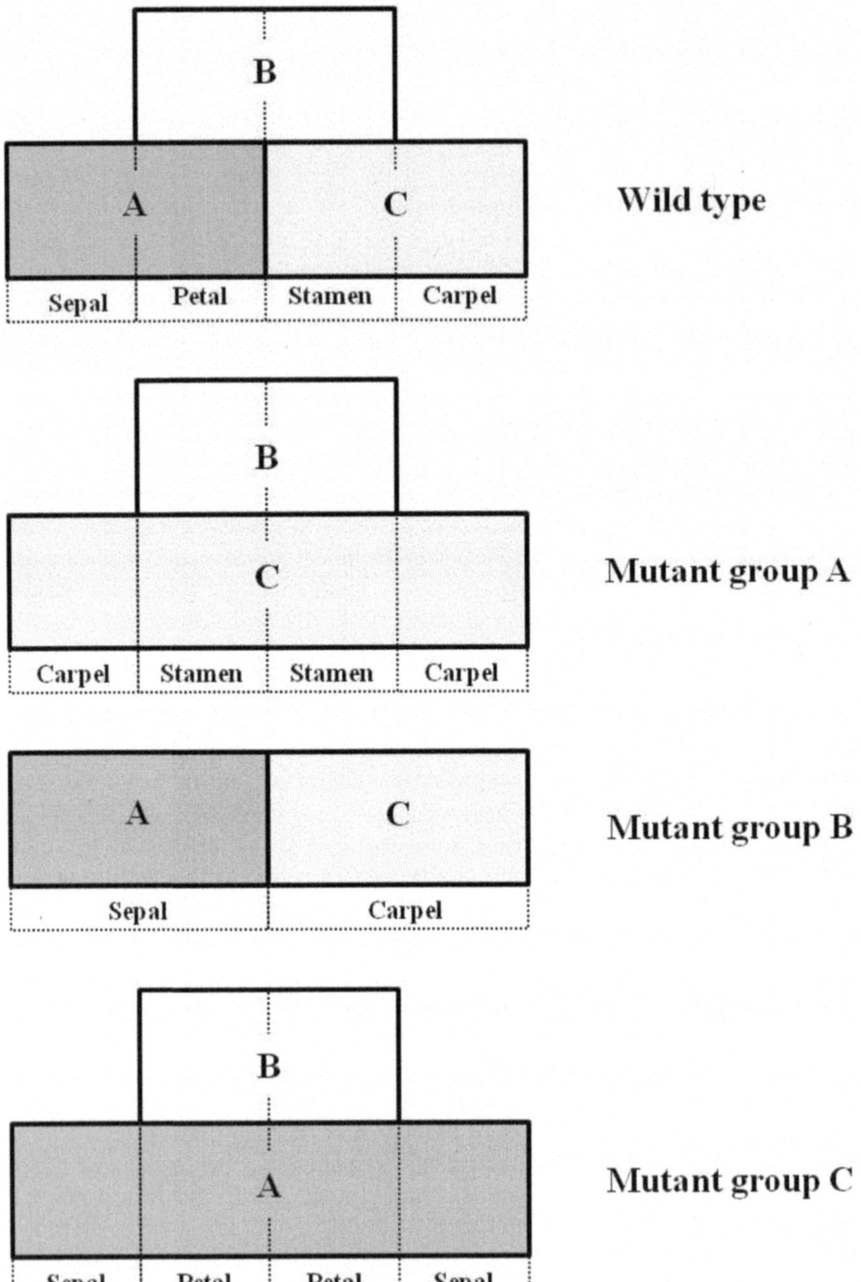

Fig. 2.19 Scheme of ABC model (Drawn by MMG Karasawa)

These observations indicate that the leaf would be the "basal state" upon which the identity of each floral organ would be determined. Based on these results, the equivalence of flowers and branches (and, consequently, of floral organs and leaves), proposed by Goethe more than 300 years ago, has been at last demonstrated (Dornelas and Dornelas 2005).

All the genes of the ABC Model code for transcription factors of the MADS family (except for *AP2*, which belongs to another transcription factor family; see Dornelas and Dornelas 2005).

In contrast to what occurs in angiosperms, our understanding of the molecular processes governing the reproductive development of gymnosperms is much limited. Genes coding transcription factors of the MADS family, which are expressed in the reproductive organs of gymnosperms, were isolated in *Pinus* (Mouradov et al. 1999). These genes showed a high similarity with angiosperm genes, highlighting the evolutionary conservation of their biological role. However, while in angiosperms these genes are responsible for the formation of sepals, petals, anthers and carpels, their role in gymnosperms is unknown. By the same token, homologues to gene *LFY* of *Arabidopsis*, widely conserved in angiosperms (Dornelas and Rodriguez 2005a, 2006) have been isolated and characterized in gymnosperms (Mouradov et al. 1998; Mellerowicz et al. 1998; Dornelas and Rodriguez 2005b). However, while in the angiosperm genomes there is only one copy of *LFY*, in the gymnosperm genomes there are two types of homologues of the gene *LFY*: the *NEEDLY*-like (*NLY*) and the *LFY*-like (*LFY*). There is evidence that the lineage that originated the angiosperms lost the homologue correspondent to NLY during its evolution (Frohlich and Parker 2000). The expression patterns of *NLY* and *LFY* in reproductive meristems of *Pinus* are similar to those observed in their angiosperm homologues, which suggests an evolutionary conservation of the function of this key element in the initiation of the reproductive development (Mouradov et al. 1998; Dornelas and Rodriguez 2005b). Although the analyses of the deduced amino acid sequences of the proteins LFY and NLY showed that they have a structure slightly different from their angiosperm homologues, transgenic *Arabidopsis* plants that superexpress *NLY* displayed early flowering when compared to non-transgenic controls (Mouradov et al. 1998). In addition, *Arabidopsis lfy* mutants showed complementation by the gene of *Pinus* (Dornelas and Rodriguez 2005b). These observations indicate that the *LFY* homologues of *Pinus* behave in a manner similar to the endogenous *LFY* gene of *Arabidopsis*, acting in a regulatory network responsible for the switch to the reproductive phase and thus showing the evolutionary conservation of the molecular mechanisms of reproductive control in plants.

2.3.6 Systems Promoting Allogamy

Plant species have a great number of reproductive strategies and systems, as shown at the end of Chap. 1. Plant sexual reproductive systems can be divided in three major classes as to the crossing mode: allogamous, autogamous and mixed (Fig. 2.20).

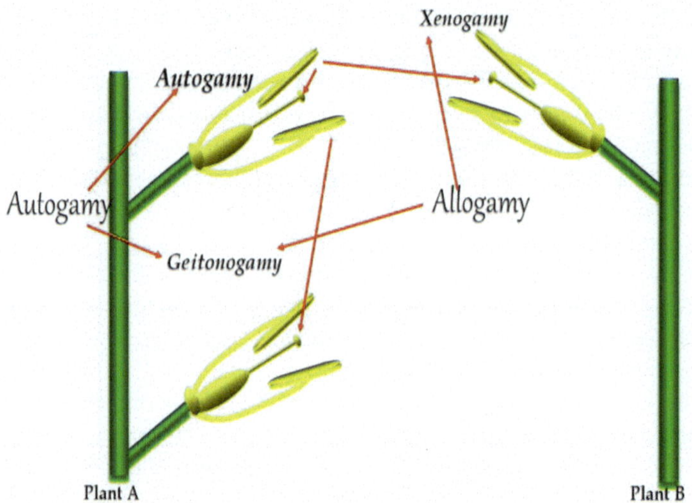

Fig. 2.20 Sexual reproductive systems (Drawn by GCX Oliveira)

A species is considered allogamous if it presents a crossing rate over 95 %. By and large, perennial species, which include most arboreal species and many crops, are allogamous. In order to assure that the plants will cross, several mechanisms can be utilized, such as self-incompatibility chemical systems, physical separation by means of unisexuality and temporal separation between pollen and stigma maturation.

- **Chemical systems: Self-Incompatibility (SI)**

Sexual incompatibility is accepted as one of the most widespread and evolutionarily most successful systems among angiosperms for promoting crossing and avoiding endogamy (Takayama and Isogai 2005; Newbigin et al. 1994). In plants, the pollen grains (male gametophytes) are transferred to the stigmas after anther dehiscence through a process known as pollination. When they get in contact with the stigma, the pollen grains absorb the water present in the superficial cells and germinate directly afterwards (Raven et al. 2007). The pollen tube grows all along the extension of the style until it penetrates the micropyle, entering the embryo sac (female gametophyte) before the double fertilization. In most angiosperms, flowers are endowed with self-incompatibility mechanisms able to recognize, hamper, or even prevent the pollen grains from fertilizing the egg cells of the same plant (Zanettini 2003). These mechanisms are controlled genetically by multiallelic loci (these loci are, indeed, among the most striking examples of multiallelism in plants) and act through the chemical interaction between the pollen and the pistil.

In relation to floral morphology, the self-incompatibility systems can be classified in two types: homomorphic (i.e., all the individuals produce the same morphology, with anthers and stigmas at the same height) and heteromorphic (i.e., there are two or three morphological types in relation to the relative height of anthers and stigmas).

In the heteromorphic types, reproductive success depends on the frequency of pollination between types with contrasting anther and stigma relative heights (Kao and Tsukamoto 2004).

The SI phenomenon has been described in many phanerogamic families of economic importance, such as Rosaceae (plum, apple), Brassicaceae (cabbage, broccoli), Fabaceae (crotalaria), Poaceae (rye), Sterculiaceae (cocoa), Passifloraceae (passion-fruit) and Solanaceae (tobacco), among others (Ramalho et al. 2004; Bueno et al. 2006). In these species, SI is controlled by a polymorphic locus called S (from self-incompatibility) that may have more than 40 alleles in natural populations (Zanettini 2003). In grasses, SI has been reported in at least 16 genera and in some species is controlled by two loci, S and Z (Baumann et al. 2000).

- **Gametophytic Self-Incompatibility**

In the gametophytic SI system the pollen-pistil interaction is determined by the haploid genome of the pollen grain and the diploid genome of the pistil (Ramalho et al. 2004; Takayama and Isogai 2005). The growth of the pollen tube is usually arrested within the style (Fig. 2.21) as a result of the contact of the mucilage secreted by the transmitting tract (Zanettini 2003). In this type of SI, the alleles exhibit a codominant interaction (Bueno et al. 2006). Each *S* allele is responsible for the production of a specific glycoprotein which can interact with others in a manner analogous to the antigen-antibody reactions in the animals, according to some authors. Thus, the glycoprotein present in the pollen is considered an "antigen" and the glycoprotein present in the stigma would be the "antibody". A male parent with genotype S_1S_2, for instance, produces pollen grains S_1 and S_2. If the female parent in the crossing is also S_1S_2, there will be no pollen tube growth, because both parents produce glycoproteins S_1 and S_2 with their respective "antigens" and "antibodies" (Schifino-Wittmann and Dall'Angol 2002). In this case, the pollen incompatibility is total. On the other hand, there can be partial compatibility when part of the pollen has no phenotypic similarity to the female parent (Fig. 2.21; Table 2.1).

- **Sporophytic Self-Incompatibility**

Sporophytic SI is determined by the diploid genotype of the microspore mother cell (Newbigin et al. 1994; Takayama and Isogai 2005) (Fig. 2.22) rather than the S allele present in the pollen (Ramalho et al. 2004; Schifino-Wittmann and Dall'Angol 2002). During microsporogenesis, the glycoproteins are produced prior to the beginning of meiosis, so that all the pollen grains receive them. Unlike the gametophytic system, sporophytic SI takes place on the stigma surface or immediately after the penetration of the pollen tube and involves substances secreted on the surface of the stigmatic papillae (Bueno et al. 2006). A kind of interaction frequently observed in sporophytic SI is complete dominance ($S_1 > S_2 > S_3 > S_4 > \ldots$) Therefore, supposing a cross between S_1S_2 and S_2S_3, a single type of glycoprotein is produced in the

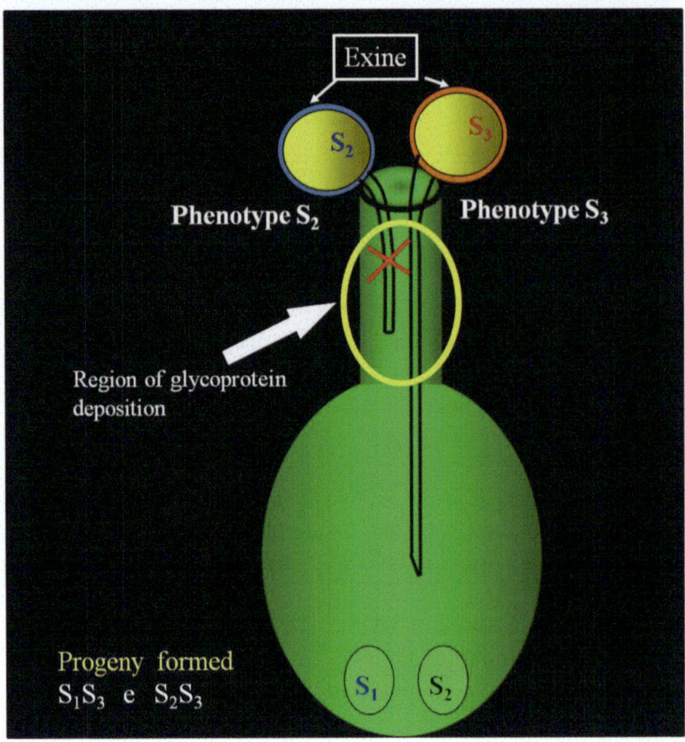

Fig. 2.21 Gametophytic self-incompatibility (Drawn by MMG Karasawa)

Table 2.1 Example of crossings between progenitors with different genotypes for the S alleles

Crossings	Pollen	Viable pollen	Progeny
S_1S_2(♀)×S_1S_2(♂)	S_1 and S_2	None	No progeny
S_1S_2(♀)×S_1S_3(♂)	S_1 and S_3	S_3	$S_1\mathbf{S_3}$ and $S_2\mathbf{S_3}$
S_1S_2(♀)×S_3S_4(♂)	S_3 and S_4	S_3 and S_4	$S_1\mathbf{S_3}$, $S_1\mathbf{S_4}$, $S_2\mathbf{S_3}$ and S2$\mathbf{S_4}$

microspore mother cell (S_1 in the first parent and S_2 in the second) and distributed to all pollen grains. Similarly, in the pistil, only the glycoprotein coded by the dominant allele is produced (again, S_1 in the first parent and S_2 in the second) (Fig. 2.22).

Consider the crossings shown below, and suppose there is dominance between the alleles:

(a) **S_1S_2(♀)×S_1S_2(♂)**
In this case, the pollen grains will have the S_1 glycoprotein. There is no growth of the pollen tube in any of the pollen grains due to the presence of glycoprotein S_1 in the female parent and, consequently, there will be no progeny.

(b) **S_1S_2(♀)×S_1S_3(♂)**
In this situation the pollen grains of the male parent will also produce the S_1 glycoprotein and we will observe the same situation as described in situation “a”

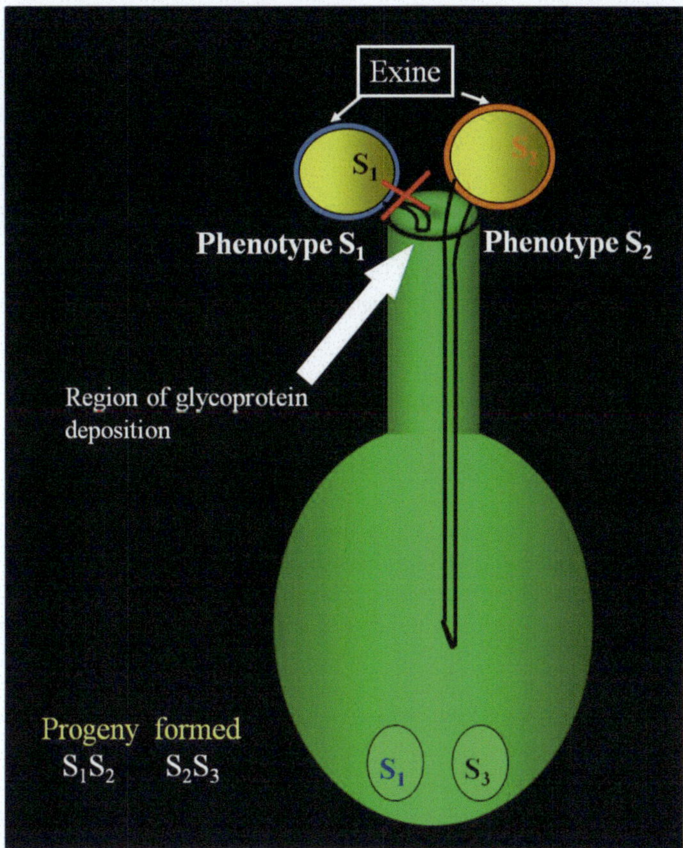

Fig. 2.22 Sporophytic self-incompatibility (Drawn by MMG Karasawa)

(c) $\mathbf{S_1S_2}$(♀) × $\mathbf{S_3S_4}$(♂)

In the situation where the male produces pollen grains containing the glycoprotein S_3, the pollen tubes will grow from all pollen grains due to the absence of the respective "antibodies" in the female and the progeny will have the genotypes S_1S_3, S_1S_4, S_2S_3 and S_2S_4.

There may be codominant interactions in the sporophytic self-incompatibility system as well. In this case, all the pollen grains receive both types of glycoproteins ("antigens") after meiosis. The pistil will also produce both types of "antibodies", as exemplified in the following crossings:

(a) $\mathbf{S_1S_2}$(♀) × $\mathbf{S_1S_2}$(♂)

Both types of pollen grain have glycoproteins S_1 and S_2 due to the codominant interaction so that no pollen tube will be formed by any pollen grain because of the presence of glycoproteins S_1 and S_2 in the female. Consequently, there will be no progeny.

(b) $\mathbf{S_1S_2}$**(♀)×**$\mathbf{S_1S_3}$**(♂)**

If glycoproteins S_1 and S_3 are produced by both types of pollen grain due to codominance no pollen tube or progeny is produced, either.

(c) $\mathbf{S_1S_2}$**(♀)×**$\mathbf{S_3S_4}$**(♂)**

In this crossing, one half of the pollen grains expresses glycoprotein S_3 and the other half expresses S4, which allows the germination of the pollen tube in both pollen grains and the generation of progeny of the type: S_1S_3, S_1S_4, S_2S_3 and S_2S_4.

Molecular Genetics of Self-Incompatibility

Research on the genic control of SI started with the observation of pistil extracts of glycoprotein-containing plants which segregated in specific S genotypes. S_2 of *Nicotiana alata* was the first such allele to be isolated. By utilizing molecular biology techniques, recent studies have shown direct evidence of the relationship between S proteins and SI systems. Two *in vivo* strategies were adopted which follow either a loss-of-function or a gain-of-function approach. In the loss-of-function experiment, an antisense S_3 cDNA (inverted sense) directed by its own promoter was introduced in S_2S_3*Petunia inflata* plants. In that way, an antisense mRNA was generated which paired to the normal endogenous mRNA inhibiting the expression of the gene in question. The transgenic plants so obtained were incapable of rejecting S_3 pollen. In the gain-of-function experiment, the S_3 gene was introduced in *P. inflata* plants with genotype S_1S_2. These plants acquired the capacity of rejecting completely S_3 pollen. Based on these studies it was established that proteins S are necessary for recognizing and rejecting non-compatible pollen (Zanettini 2003). Later on, the cDNA corresponding to alleles S_2, S_3 and S_6 was sequenced and its conserved regions were identified. This information was used for cloning the cDNA of other alleles in *Nicotiana alata*, *Petunia inflata*, *P. hybrida*, *Solanum chacoense*, *S. tuberosum* and *S. peruvianum*. Sequence alignment showed that 16 % of the amino acids were conserved, including eight to ten cysteine residues. Five conserved regions containing cysteine residues were found in Solanaceae. Histidine residues were also found in two of them. Later use of Southern Blot in the analyses of individuals homozygotic for S_1, S_2, S_3, S_6 and S_7 demonstrated that genic control was done by a single locus (Newbigin et al. 1994) with many alleles (Kao and McCubbin 1996).

Gene control: Gametophytic System

In grasses, several species are described where the control of SI is done by two non-linked loci (S and Z), which have already been added to linkage maps of rye, oats and barley. Based on the mutants isolated from grasses, at least four genes are believed to be involved in the control of SI, of which two are present in the pollen and two in the stigma (Baumann et al. 2000).

A model for the gene control of the gametophytic system has been described for Solanaceae, Rosaceae and Scrophulariaceae. Locus *S* is composed of two genes,

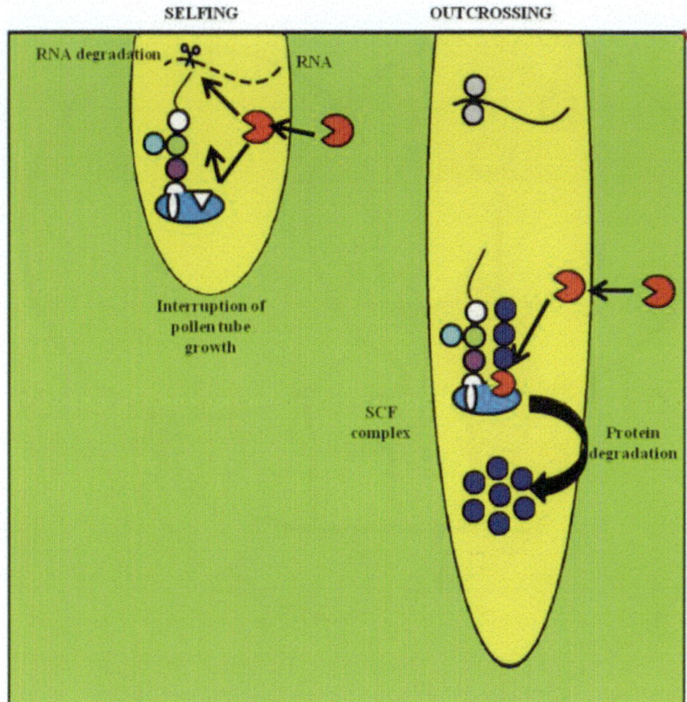

Fig. 2.23 Scheme of genetic control of *Solanaceae*, *Rosaceae* and *Scrophulariaceae* (Drawn and modified by MMG Karasawa based on Takayama and Isogai 2005)

S-RNase and *SLF/SFB* (Fig. 2.23). S-RNase is secreted in large amounts to the extracellular matrix of the style, whence it is transferred to the pollen tube, where it works as a cytotoxin that degrades RNA.

However, degradation only occurs with self-pollen. *SLF/SFB* are male determinant genes with a motif belonging to the F-box protein family which usually acts as ubiquitin ligand components and are expected to be involved in the degradation of proteins (Takayama and Isogai 2005).

In Papaveraceae, the only female determinant induces the increase in the concentration of Ca^{2+} during the initial 10-min interaction of the incompatible pollen (Fig. 2.24). The key mechanism in the growth inhibition of the incompatible pollen tube consists in the influx of Ca^{2+} into incompatible pollen, which results in a rapid depolymerization of the actin filaments for 60 s, sustained for about an hour. Afterwards, a phosphorylation of soluble pyrophosphatases occurs. Both calcium and phosphorylation inhibit pyrophosphatase activity, leading to a reduction in the biosynthetic efficiency of pollen and to the inhibition of pollen tube growth, ending up in its death (Takayama and Isogai 2005). During this period, dramatic alterations in mitochondria, Golgi apparatus and endoplasmic reticulum morphology are observed, and some of these organelles are completely degraded (Bosch and Franklin-Tong 2008).

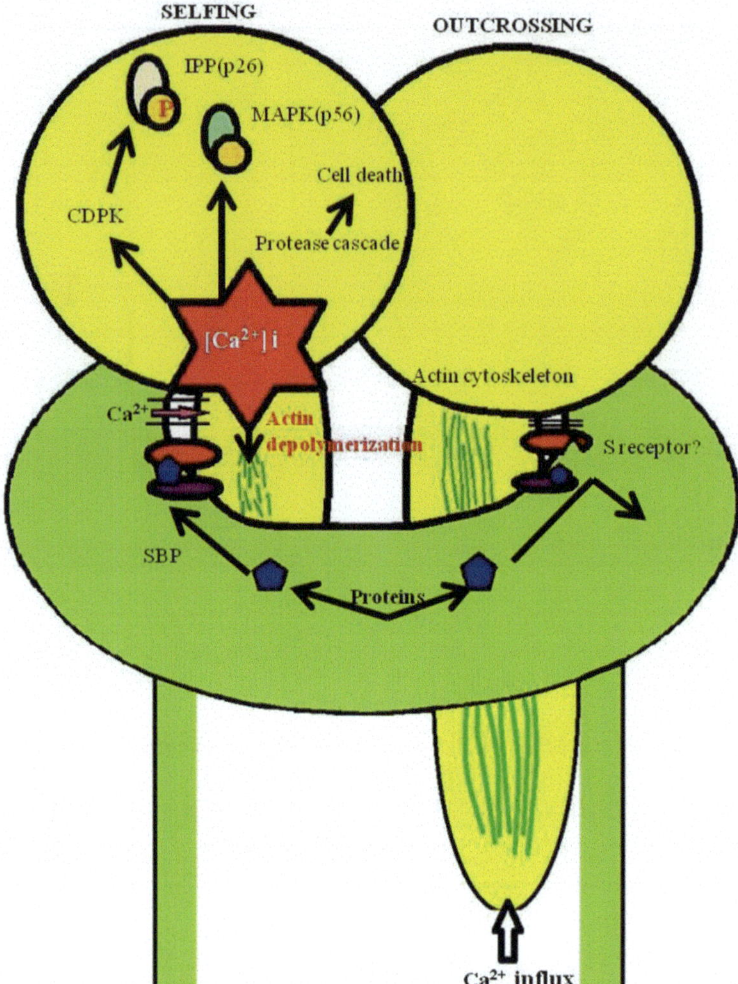

Fig. 2.24 Scheme of gene control of *Papaveraceae* (Drawn and modified by MMG Karasawa based on Takayama and Isogai 2005)

Gene Control: Sporophytic System

A gene control model for the sporophytic system has been defined for brassicas. In these plants, the SI control locus consists of three genes: *SP11*, *SRK* and *SLG* (Fig. 2.25). The gene *SRK*, which plays a role as a stigmatic kinase receptor, is the female determinant located in the plasma membrane of the papilla cells; *SP11*, the male determinant, is predominantly expressed in the anther tapetum and accumulated in the chambers of the columella, in the pollen grain wall. After pollination, SP11 penetrates the walls of the papilla cells and attaches to SRK in a specific

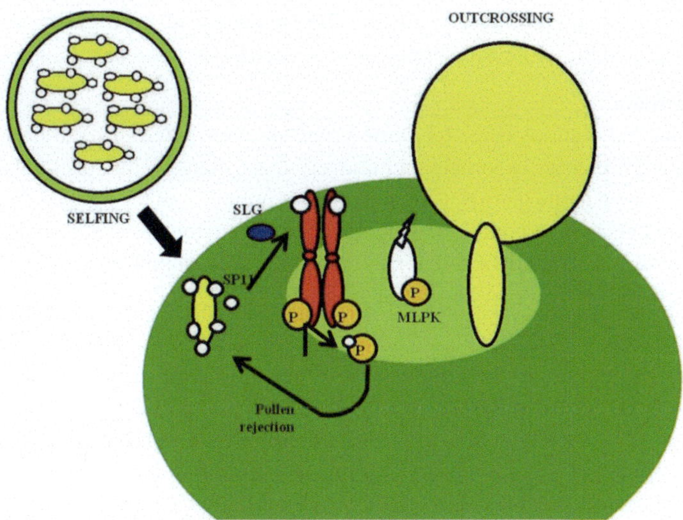

Fig. 2.25 Scheme of gene control in *Brassicaceae* (Drawn and modified by MMG Karasawa based on Takayama and Isogai 2005)

S-haplotype manner. This attachment induces the autophosphorylation of SRK, followed by a signal cascade that results in the rejection of the incompatible pollen. The gene *SLG* is not essential for recognition or rejection, but is situated in the papilla cells, potentializing the SI reaction of some S haplotypes (Takayama and Isogai 2005).

Break of the Self-Incompatibility Control

The control of SI can be broken by three factors: duplication of locus *S*, mutations that provoke the loss of activity of S-RNase, and mutations that do not provoke the loss of activity of S-RNase (Stone 2002). In Fabaceae, Onagraceae and Rosaceae, the loss-of-function seems to be simple, but in Solanaceae the mutations are more complex and are frequently associated to the duplication of an S allele, which may indicate some gain-of-function (Golz et al. 2000).

- **Physical systems**
  - **Unisexuality**
    - **Dicliny**

According to Richards (1997), any population where the members are not regularly hermaphroditic are considered diclinous (Fig. 1.20). Many conditions affect the distribution of the sexes in a population. These sex distribution patterns are believed to have evolved for preventing or reducing the frequency of selfing, which leads to endogamy and to the expression of deleterious alleles (for more details, see Chap. 1).

- **Dioecy**

In dioecious populations (Fig. 1.22) the sexes are distributed in separate individuals, i.e., there are entirely female and entirely male plants. This mechanism is not much common in angiosperms, and is present in only 4 % of the species (Fig. 1.21). Dioecy prevents selfing completely and promotes crossing; however, it seems inefficient, because only half of the fertile branches in a population produces seeds (Richards 1997).

- **Monoecy**

Monoecious populations have flowers with separate sexes allocated in different parts of the plant or the inflorescence (Fig. 1.22). This mechanism is found in 7 % of the angiosperms (Fig. 1.21). Monoecy is supposed to have evolved because it prevents selfing, but it is not much efficient, since male flowers can pollinize female flowers in the same plant (geitonogamy; for details, see Chap. 1).

- **Heterostyly or Heteromorphism**

Heterostylous populations have different flower morphotypes among individuals; each individual has only one flower morphotype. The morphotypes differ in the relative heights of anthers and stigmas; according to the species, there may be two or three relative height patterns (Fig. 2.26). This mechanism also reduces the incidence of selfing.

- **Temporal systems**
  - **Protandry**

The function of this system is to promote allogamy through the anticipated maturation of pollen in relation to the stigma of the same flower, while allowing the fertilization of other flowers present in different parts of the same individual (geitonogamy) or of other individuals (xenogamy), provided their stigmas are mature.

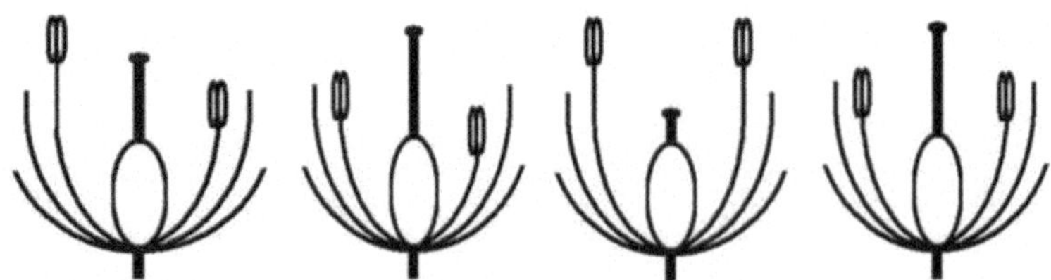

Fig. 2.26 Schematic drawing of heteromorphic forms (drawn by MMG Karasawa)

- **Protogyny**

Protogyny consists in the anticipated maturation of the stigma in relation to pollen, which makes it obligatory that each flower be pollinated by geitonogamy or xenogamy.

2.3.7 Autogamy

A species is autogamous, by convention, if its crossing rate is lower than 5 % or, in other words, its selfing rate is between 95 and 100 % inclusive. The flowers of autogamous species are usually small and white or have colors unattractive to pollinators. Goodwillie et al. (2005) evaluated 345 species and found this condition to be relatively rare in nature. Only 10 % of the species studied showed this form of reproduction.

2.3.8 Mixed Mating System

All the species that reproduce by utilizing both allogamy and autogamy simultaneously are considered mixed species. Their crossing rates range from 5 to 95 %, depending on the environmental conditions and the pollinator frequency. Goodwillie et al. (2005) suggest that the mixed system is common, comprising about 80 % plus of the angiosperms and gymnosperms.

Bibliography

Albertini E, Barcaccia G, Porceddu A, Sorbolini S, Falcinelli M (2001a) Mode of reproduction is detected by Parth1 and Sex1 SCAR markers in a wide range of facultative apomictic Kentucky bluegrass varieties. Mol Breed 7:293–300

Albertini E, Porceddu A, Ferranti F, Reale L, Barcaccia G, Romano B, Falcinelli M (2001b) Apospory and parthenogenesis may be uncoupled in *Poa pratensis*: a cytological investigation. Sex Plant Reprod 14:213–217

Albertini E, Marconi G, Barcaccia G, Raggi L, Falcinelli M (2004) Isolation of candidate genes for apomixis in *Poa pratensis* L. Plant Mol Biol 56(6):879–894

Alves ER, Carneiro VTC, Araújo ACG (2001) Direct evidence of pseudogamy in an apomictic *Brachiaria brizantha* (Poaceae). Sex Plant Reprod 14(4):207–212

Appels R, Morris R, Gill BK, May CE (1998) Chromosome biology. Kluwer Academic, Norwell, cap. 10 e 12

Araújo ACG, Mukhambetzhanov S, Pozzobon MT, Santana EF, Carneiro VTC (2000) Female gametophyte development in apomictic and sexual *Brachiaria brizantha* (Poaceae). Revue de Cytologie et de Biologie Vegetales–Le Botaniste Tome 23:13–28

Araújo ACG, Falcão R, Simões KCR, Carneiro VTC (2004) Identificação de acessos de *Brachiaria* com interesse ao estudo da apomixia facultativa, Boletim de Pesquisa e desenvolvimento 74. Embrapa Recursos Genéticos e Biotecnologia, Brasília

Araujo ACG, Nóbrega JM, Pozzobon MT, Carneiro VTC (2005) Evidence of sexuality in induced tetraploids of *Brachiaria brizantha* (Poaceae). Euphytica 144:39–50

Asker S (1980) Gametophytic apomixis: elements and genetic regulation. Hereditas 93(2): 277–293

Asker SE, Jeling L (1992) Apomixis in plants. CRC, Boca Raton, 298 p

Barcaccia G, Arzenton F, Sharbel TF, Varotto S, Parrini P, Lucchin M (2006) Genetic diversity and reproductive biology in ecotypes of the facultative apomict *Hypericum perforatum* L. Heredity 96(4):322–334

Bashaw EC (1980) Apomixis and its application in crop improvement. In: Fehr WR, Hadley HH (eds) Hybridization of crop plants. American Society of Agronomy/Crop Science Society of America, Madison, pp 45–63

Baumann U, Juttner J, Bian X, Langridge P (2000) Self-incompatibility in the grasses. Ann Bot 85:203–209

Bicknell RA, Koltunow AM (2004) Understanding apomixis: recent advances and remaining conundrums. Plant Cell 16(Suppl Plant Reproduction):S228–S245

Bicknell R, Borst NK, Koltunow AM (2000) Monogenic inheritance of apomixis in two *Hieracium* species with distinct developmental mechanisms. Heredity 84:228–237

Bicknell R, Podivinsky E, Catanach A, Erasmuson S, Lambie S (2001) Strategies for isolating mutants in *Hieracium* with dysfunctional apomixis. Sex Plant Reprod 14:227–232

Blázquez MA, Soowal LN, Lee I, Weigel D (1997) LEAFY expression and flower initiation in Arabidopsis. Development 124:3835–3844

Bonilla JR, Quarin CL (1997) Diplosporous and aposporous apomixis in a pentaploid race of *Paspalum minus*. Plant Sci 127:97–104

Bosch M, Franklin-Tong VE (2008) Self-incompatibility in Papaver: signaling to trigger PCD in incompatible pollen. J Exp Bot 59(3):481–490

Brown WV, Emery WHP (1958) Apomixis in the *Gramineae*: Panicoideae. Am J Bot 45:253–263

Carman JG (1997) Asynchronous expression of duplicate genes in angiosperms may cause apomixis, bispory, tetraspory, and polyembryony. Biol J Linn Soc 61:51–94

Carman JG (2001) The gene effect: genome collision and apomixis. In: Savidan Y, Carman JG, Dresselhaus T (eds) The flowering of apomixis: from mechanisms to genetic engineering. CIMMYT, IRD, European Commission DG (VI), Mexico, pp 95–110

Catanach AS, Erasmuson SK, Podivinsky E, Jordan BR, Bicknell R (2006) Deletion mapping of genetic regions associated with apomixis in Hieracium. Proc Natl Acad Sci U S A 103(49):18650–18655, www.pnas.org/cgi/doi/10.1073/pnas.0605588103

Cavalli SS (2003) Apomixia: um método de reprodução assexual. In: de Freitas LB, Bered F (eds) Genética e evolução vegetal. UFRGS, Porto Alegre, pp 41–55

Chaudhury AM, Koltunow A, Payne T, Luo M, Tucker MR, Dennis ES, Peacock WJ (2001) Control of early seed development. Annu Rev Cell Dev Biol 17:677–699

Coen ES, Meyerowitz EM (1991) The war of the whorls: genetic interactions controlling flower development. Nature 353:31–37

Curtis MD, Grossniklaus U (2008) Molecular controlo f autonomous embryo and endosperm development. Sex Plant Reprod 21:79–88

Czapik R (1994) How to detect apomixis in *Angiospermae*. Polish Bot Stud 8:13–21

Bueno LC de S, Mendes ANG, Carvalho SP (2006) Melhoramento genético de plantas: princípios e procedimentos, 2ªth edn. UFLA, Lavras, 319 p

Dornelas MC (2003) Desenvolvimento e diferenciação celular. In: Freitas LB de, Bered F (eds) Genétsica e evolução vegetal, pp 87–110

Dornelas MC, Dornelas O (2005) From leaf to flower: revisiting Goethe's concepts on the "metamorphosis" of plants. Braz J Plant Physiol 17:335–343

Dornelas MC, Rodriguez APM (2005a) The rubber tree (Hevea brasiliensis Muell. Arg.) homologue of the LEAFY/FLORICAULA gene is preferentially expressed in both male and female floral meristems. J Exp Bot 56:1965–1974

Dornelas MC, Rodriguez APM (2005b) A FLORICAULA/LEAFY gene homolog is preferentially expressed in developing female cones of the tropical pine Pinus caribaea var. caribaea. Genet Mol Biol 28:299–307

Dornelas MC, Rodriguez APM (2006) The tropical cedar tree (Cedrela fissilis, Meliaceae) homolog of the Arabidopsis LEAFY gene is expressed in reproductive tissues and can complement Arabidopsis leafy mutants. Planta 223:306–314

Drews GN, Yadegary R (2002) Development and function of the angiosperm female gametophyte. Annu Rev Genet 36:99–124

Dusi DMA (2001) Apomixis in *Brachiaria decumbens* Stapf. PhD Thesis, University of Wageningen, The Netherlands

Dusi DMA, Willemse MTM (1999) Apomixis in *Brachiaria decumbens* Stapf.: gametophytic development and reproductive calendar. Acta Botanica Cracoviensia (Series Botanica) 41:151–162

Estrada-Luna AA, Huanca-Mamani W, Acosta-Garcia G, Leon-Martinez G, Becerra-Flora A, Pérez-Ruíz R, Vielle-Calzada J-PH (2002) Beyond promiscuity: from sexuality to apomixis in flowering plants. In Vitro Cell Dev Biol Plant 38:146–151

Frohlich MW, Parker DS (2000) The mostly male theory of flower evolutionary origins. Syst Bot 25:155–170

Fryxel P (1957) Mode of reproduction of higher plants. Bot Rev 23:135–233

Goldberg RB, Beals TP, Sanders PM (1993) Anther development: basic principles and practical applications. Plant Cell 5:1217–1229

Golz JF, Clarke AE, Newbigin E (2000) Mutational approaches to the study of self-incompatibility: revisiting the pollen-part mutants. Ann Bot 85(Suppl A):95–103

Goodwillie C, Kalisz S, Eckert CG (2005) The evolutionary enigma of mixed mating systems in plants: occurrence, theoretical explanations, and empirical evidence. Annu Rev Ecol Evol Syst 36:47–79

Grossniklaus U, Vielle-Calzada J-P, Hoeppner M, Gagliano W (1998) Maternal control of embryogenesis by MEDEA, a Polycomb-group gene in *Arabidopsis*. Science 280:446–450

Grossniklaus U, Spillane C, Page DR, Köhler C (2001) Genomic imprinting and seed development: endosperm formation with and without sex. Curr Opin Plant Biol 4:21–27

Hanna WW, Bashaw EC (1987) Apomixis: its identification and use in plant breeding. Crop Sci 27:1136–1139

Hartmann HD, Kester DE (1975) Propagación de plantas: principios y prácticas, 4th edn. Compañia editorial continental, México, 809 p

Huck N, Moore JM, Federer M, Grossniklaus U (2003) The *Arabidopsis* mutant feronia disrupts the female gametophytic control of pollen tube reception. Development 130:2149–2159

Kao T-H, McCubbin AG (1996) How flowering plants discriminate between self and non-self pollen to prevent inbreeding. Proc Natl Acad Sci U S A 93:12059–12065

Kao T-H, Tsukamoto T (2004) The molecular and genetic bases of S-RNase-based self-incompatibility. Plant Cell 16(Suppl):S72–S83, www.aspb.org, www.plantcell.org/cgi/doi/10.10.1105/tpc.016154

Karasawa MMG (2005) Análise da estrutura genética de populações e sistema reprodutivo de *Oryza glumaepatula* por meio de microssatélites. Tese (Doutorado)—Escola Superior de Agricultura "Luiz de Queiroz", Universidade de São Paulo, Piracicaba, 91 p

Karasawa MMG, Oliveira GCX, Veasey EA, Vencovsky R (2006) Evolução das plantas e de sua forma de reprodução. M.M.G.Karasawa, Piracicaba, 86 p

Kaushal P, Malaviya DR, Roy AK, Shalini P, Agrawal A, Ambica K, Siddiqui SA (2008) Reproductive pathways of seed development in apomictic guinea grass (*Panicum maximum* Jacq.) reveal uncoupling of apomixis components. Euphytica 164:81–92. doi:10.1007/s10681-008-9650-4-1573-5060

Khush GS (1994) Apomixis: exploiting hybrid vigor in rice. IRRI, Los Banos

Kinoshita T, Yadegari R, Harada JJ, Goldberg RB, Fischer RL (1999) Imprinting of the MEDEA Polycomb gene in the *Arabidopsis* endosperm. Plant Cell 11:1945–1952

Knox RB (1967) Apomixis: seasonal and population differences in grass. Science 157:325–326

Koltunow AM (1993) Apomixis: embryo sacs and embryo formed without meiosis or fertilization in ovules. Plant Cell 5:1425–1437

Koltunow AM, Grossniklaus U (2003) Apomixis: a developmental perspective. Annu Rev Plant Biol 54:547–574

Koltunow AM, Johnson SD, Bicknell RA (1998) Sexual and apomictic development in *Hieracium*. Sex Plant Reprod 11:213–230

Koltunow AM, Johnson SD, Bicknell RA (2000) Apomixis is not developmentally conserved in related, genetically characterized *Hieracium* plants of varying ploidy. Sex Plant Reprod 12:253–266

Lakshmanan KK, Ambegaokar KB (1984) Polyembryony. In: Johri BM (ed) Embryology of *Angiosperms*. Springer, Berlin, pp 445–474

Leblanc O, Peel MD, Carman JG, Savidan Y (1995a) Megasporogenesis and mega-gametogenesis in several *Tripsacum* species (Poaceae). Am J Bot 82:57–63

Leblanc O, Grimanelli D, González-de-León D, Savidan Y (1995b) Detection of the apomictic mode of reproduction in maize-*Tripsacum* hybrids using maize RFLP markers. Theor Appl Genet 90:1198–1203

Luo M, Bilodeau P, Koltunow AM, Dennis ES, Peacock WJ, Chaudhury AM (1999) Genes controlling fertilization-independent seed development in *Arabidopsis* thaliana. Proc Natl Acad Sci U S A 96:296–301

Luo M, Bilodeau P, Dennis ES, Peacock WJ, Chaudhury AM (2000) Expression and parent-of-origin effects for FIS2, MEA, and FIE in the endosperm and embryo of developing Arabidopsis seeds. Proc Natl Acad Sci U S A 97:10637–10642

Ma H (1998) To be, or not to be, a flower: control of floral meristem identity. Trends Genet 14:26–32

Ma H (2005) Molecular genetic analyses of microsporogênesis and microgametogênesis in flowering plants. Annu Rev Plant Biol 56:303–434

Mandel MA, Gustafson-Brown C, Savidge B, Yanofsky MF (1992) Molecular characterization of the Arabidopsis floral homeotic gene APETALA1. Nature 360:273–277

Martínez EJH, Hopp E, Stein J, Ortiz JPA, Quarin CL (2003) Genetic characterization of apospory in tetraploid *Paspalum notatum* based on the identification of linked molecular markers. Mol Breed 12(4):319–327

Martínez EJ, Acuña CA, Hojsgaard DH, Tcach MA, Quarin CL (2007) Segregation for sexual seed production in *Paspalum* as directed by male gametes of apomictic triploid plants. Ann Bot 100(6):1239–1247

Márton I., Dresselhaus T. 2008. A comparison of early molecular fertilization mechanisms in animals and flowering plants. Sexual Plant Reproduction, 21(1):37–52

Matzk F, Meister A, Brutovska R, Schubert I (2001) Reconstruction of reproductive diversity in *Hypericum perforatum* L. opens novel strategies to manage apomixis. Plant J 26:275–282

Matzk F, Prodanovic S, Bäumlein H, Schubert I (2005) The inheritance of apomixis in *Poa pratensis* confirms a five locus model with differences in gene expressivity and penetrance. Plant Cell 17(1):13–24

Matzk, F. 2007. Flow Cytometry with Plant Cells: Analysis of Genes, Chromosomes and Genomes. Copyright © 2007 Wiley-VCH Verlag GmbH & Co. KGaA. doi:10.1002/9783527610921

Maunseth JD (1995) Botany: an introduction to plant biology. Saunders, Philadelphia

Maynard Smith J (1971) The origin and maintenance of sex. In: Williams GC (ed) Group selection. Aldine Atherton, Chicago, pp 163–175

Mellerowicz EJ, Horgan K, Walden A, Coker A, Walter C (1998) PRFLL, a *Pinus radiata* homologue of FLORICAULA and LEAFY is expressed in buds containing vegetative shoot and undifferentiated male cone primordia. Planta 206:619–629

Meyerowitz EM, Bowman JL, Brockman LL, Drews GL, Jack T, Sieburth LE, Weigel D (1991) A genetic and molecular model for flower development in *Arabidopsis thaliana*. Development 112(Suppl):157–169

Miles J. W., Escandón M. L. 1997 Further evidence on the inheritance of reproductive mode in Brachiaria . Canadian Journal of Plant Science, 77(1):105–107. doi:10.4141/P95-187

Mouradov A, Glassick T, Hamdorf B, Murphy L, Fowler B, Marla S, Teasdale RD (1998) NEEDLY, a *Pinus radiata* ortholog of FLORICAULA/LEAFY genes, expressed in both reproductive and vegetative meristems. Proc Natl Acad Sci U S A 95:6537–6542

Mouradov A., Hamdorf B., Teasdale R.D., Kim J.T., Winter, K-U, Theißen G. 1999. DEF/GLO-like MADS-box gene from a gymnosperm: Pinus radiata contains an ortholog of angiosperm B class floral homeotic genes. Developmental Genetics - Special Issue: Focus on Floral Development, 25 (3): 245–252. doi:10.1002/(SICI)1520-6408. Copyright © 1999 Wiley-Liss, Inc.

Newbigin ED, Anderson MA, Clarke AE (1994) Gametophytic self-incompatibility in *Nicotiana alata*. In: Williams EG, Clarke AE, Knox RB (eds) Genetic control of self-incompatibility and reproductive development in flowering plants, 2nd edn. Kluwer Academic, Dordrecht, pp 5–18

Nogler GA (1984) Gametophytic apomixis. In: Johri BM (ed) Embryology of angiosperms. Springer, Berlin, pp 475–518

Noyes RD, Baker R, Mai B (2007) Mendelian segregation for two-factor apomixis in Erigeron annuus (*Asteraceae*). Heredity 98:92–98

Ozias-Akins P, van Dijk PJ (2007) Mendelian genetics of apomixis in plants. Annu Rev Genet 41:509–537

Ozias-Akins P, Roche D, Hanna WW (1998) Tight clustering and hemizygosity of apomixis-linked molecular markers in Pennisetum squamulatum implies genetic control of apospory by a divergent locus that may have no allelic form in sexual genotypes. Proc Natl Acad Sci U S A 95:5127–5132

Peña L, Martín-Trillo M, Juárez J, Piña JA, Navarro L, Martinez-Zapater JM (2001) Constitutive expression of Arabidopsis LEAFY or APETALA1 genes in citrus reduces their generation time. Nat Biotechnol 19:215–216

Porceddu A, Albertini E, Barcaccia G, Marconi G, Bertoli F, Veronesi F (2002) Linkage mapping in apomictic and sexual Kentucky bluegrass (*Poa pratensis* L.) genotypes using a two way pseudotestcross strategy based on AFLP and SAMPL markers. Theor Appl Genet 104:273–280

Punwani JA, Drews GM (2008) Development and function of synergid cell. Sex Plant Reprod 21:7–15

Pupilli F, Labombarda P, Caceres ME, Quarin QL, Arcioni S (2001) The chromosome segment related to apomixis in Paspalum simplex is homoeologous to the telomeric region of the long arm of rice chromosome 12. Mol Breed 8(1):53–61

Pupilli F, Martinez EJ, Busti A, Calderini O, Quarin CL, Arcioni S (2004) Comparative mapping reveals partial conservation of synteny at the apomixis locus in *Paspalum* spp. Mol Genet Genomics 270(6):539–548

Ramalho MAP, dos Santos JB, Pinto CABP (2004) Genética na agropecuária, 3rd edn. UFLA, Lavras, 472 p

Raven PH, Evert RF, Eichhorn SE (2007) Biologia vegetal, 7th edn. Guanabara Koogan, Rio de Janeiro, 830 p

Richards AJ (1997) Plant breeding systems, 2nd edn. Chapman & Hall, Cambridge, 529 p

Roche D, Cong P, Chen Z, Hanna WW, Gustine DL, Sherwood RT, Ozias-Akins P (1999) An apospory-specific genomic region is conserved between buffelgrass (*Cenchrus ciliaris* L.) and *Pennisetum squamulatum* Fresen. Plant J 19:203–208

Rodrigues JCM, Koltunow AMG (2005) Epigenetic aspects of sexual and asexual seed development. Acta Biol Cracov Bot 47:37–49

Rodrigues JCM, Cabral GB, Dusi DMA, Mello LV, Ridgen D, Carneiro VTC (2003) Identification of differentially expressed cDNA sequences in ovaries of sexual and apomictic plants of *Brachiaria brizantha*. Plant Mol Biol 53:745–757

Savidan Y (1989) Another working hypothesis for the control of parthenogenesis in *Panicum*. Apomixis Newsletter 1:47–51

Savidan Y (2000) Apomixis: genetics and breeding. Plant Breed Rev 18:13–86

Savidan Y (2001) Transfer of apomixis through wide crosses. In: Savidan Y, Carman JG, Dresselhaus T (eds) The flowering of apomixis: from mechanisms to genetic engineering. CIMMYT, IRD, European Commission DG VI (FAIR), Mexico DF, pp 153–167

Schifino-Wittmann MT, Dall'Angol M (2002) Auto-incompatibilidade em plantas. Ciência Rural 32(6):1083–1090

Sharbel TF, Mitchell-Olds T (2001) Recurrent polyploid origins and biogeographical variation in the *Arabis holboellii* complex (*Brassicaceae*). Heredity 87:59–68

Spillane C, MacDougall C, Stock C, Köhler C, Vielle-Calzada J-P, Nunes SM, Grossniklaus U, Goodrich J (2000) Interaction of the Arabidopsis Polycomb group proteins FERTILIZATION-INDEPENDENT ENDOSPERM and MEDEA mediates their common phenotypes. Curr Biol 10:1535–1538

Stein J, Pessino SC, Martínez EJ, Rodriguez MP, Siena LA, Quarin CL, Ortiz JPA (2003) A genetic map of tetraploid *Paspalum notatum* Flügge (bahiagrass) based on single-dose molecular markers. Mol Breed 20(2):153–166

Stone JL (2002) Molecular mechanisms underlying the breakdown of gametophytic self-incompatibility. Q Rev Biol 77(1):17–32

Takayama S, Isogai A (2005) Self-incompatibility in plants. Annu Rev Plant Biol 56:467–489

Tucker MR, Araújo ACG, Paech N, Hecht V, Schmidt EDL, Rossel J-B, De Vries SC, Koltunow AMG (2003) Sexual and apomictic reproduction in *Hieracium* subgenus Pilosella are closely interrelated developmental pathways. Plant Cell 15:1524–1537

Valle CB, Glienke C, Leguisamon GOC (1994) Inheritance of apomixis in *Brachiaria* a tropical forage grass. Apomixis Newsletter 7:42–43

Vielle-Calzada J-P, Thomas J, Spillane C, Coluccio A, Hoeppner MA, Grossniklaus U (1999) Maintenance of genomic imprinting at the *Arabidopsis* MEDEA locus requires zygotic DDM1 activity. Genes Dev 13:2971–2982

Weigel D., Nilsson, Ove. 1995. A developmental switch sufficient for flower initiation in diverse plants. Nature 377(6549):495-500. doi:10.1038/377495a0

Weigel D (1998) Floral induction to floral shape. Curr Opin Plant Biol 1:55–59

Yadegari R, Kinoshita T, Lotan O, Cohen G, Katz A, Choi Y, Katz A, Nakashima K, Harada JJ, Goldberg RB, Fischer RL, Ohad N (2000) Mutations in the FIE and MEA genes that encode interacting polycomb proteins cause parent-of-origin effects on seed development by distinct mechanisms. Mol Breed 20(2):153–166

Yeung EC, Meinke DW (1993) Embryogenesis in angiosperms: development of the suspensor. Plant Cell 5:1371–1381

Zanettini MHB (2003) Sistemas de incompatibilidade. In: Freitas LB, Bered F (eds) Genética e evolução vegetal. Editora da UFRGS, Porto Alegre, 463 p

Zanettini MHB, Lauxen M da S (2003) Reprodução nas angiospermas. In: Freitas LB de, Bered F (eds) Genética e evolução vegetal, pp 29–40

Záveský L, Jarolímová V, Štěpánek J (2007) Apomixis in *Taraxacum paludosum* (section Palustria, Asteraceae): recombinations of apomixis elements in inter-sectional crosses. Plant Syst Evol 265(3–4):147–163

Index

M.M. Gniech Karasawa (ed.), *Reproductive Diversity of Plants*,
DOI 10.1007/978-3-319-21254-8

The manufacturer's authorised representative in the EU is Springer Nature Customer Service Centre GmbH, Europaplatz 3, 69115 Heidelberg, Germany. If you have any concerns regarding our products, please contact ProductSafety@springernature.com

Printed and bound by CPI Group (UK) Ltd, Croydon, CR0 4YY
15/07/2026
02167654-0001